UNIPA Springer Series

The **UNIPA Springer Series** publishes single and co-authored thematic collected volumes, monographs, handbooks and advanced textbooks on specific issues of particular relevance in six core scientific areas. The issues may be interdisciplinary or within one specific area of interest. Manuscripts are invited for publication in the following fields of study:

1- Clinical Medicine;
2- Biomedical and Life Sciences;
3- Engineering and Physical Sciences;
4- Mathematics, Statistics and Computer Science;
5- Business, Economics and Law;
6- Human, Behavioral and Social Sciences.

Manuscripts submitted to the series are peer reviewed for scientific rigor followed by the usual Springer standards of editing, production, marketing and distribution. The series will allow authors to showcase their research within the context of a dynamic multidisciplinary platform. The series is open to academics from the University of Palermo but also from other universities around the world. Both scientific and teaching contributions are welcome in this series. The editorial products are addressed to researchers and students and will be published in the English language.

The volumes of the series are single-blind peer-reviewed.

More information about this series at http://www.springer.com/series/13175

Michele Ciofalo

Thermofluid Dynamics
of Turbulent Flows

Fundamentals and Modelling

UNIVERSITÀ
DEGLI STUDI
DI PALERMO

Springer

Michele Ciofalo
Department of Engineering
University of Palermo
Palermo, Italy

ISSN 2366-7516 ISSN 2366-7524 (electronic)
UNIPA Springer Series
ISBN 978-3-030-81080-1 ISBN 978-3-030-81078-8 (eBook)
https://doi.org/10.1007/978-3-030-81078-8

This Springer imprint is published by the registered company Springer Nature Switzerland AG
The registered company address is: Gewerbestrasse 11, 6330 Cham, Switzerland

To the dear memory of Michael W. Collins.

Preface

It is a riddle, wrapped in a mystery, inside an enigma
Winston Churchill

The *enigma* to which Winston Churchill hinted was the world beyond the *iron curtain* (a term he himself coined), but the definition fits well also the world beyond *laminar* that is the vast domain of *turbulence*.

Doubtless, the hundred and fifty years that separate us from the pioneering studies of Boussinesq or Reynolds have shed light on the most enigmatic aspects of fluid turbulence. The development of the mathematical theory of nonlinear dynamic systems, starting in the 1960s, has allowed the behaviour of a fluid in turbulent motion to be viewed in the larger context of dynamical systems with "strange" attractors, explaining both its chaotic, pseudo-random, nature and its sensitivity to the initial conditions. In about the same period, the increase in computational power, with the impressive trend that continues up to now, has made direct numerical simulation feasible for an increasing range of turbulent flows. Finally, the improvement of experimental techniques has allowed the acquisition of a larger and larger database on fluid turbulence under the most diverse geometrical and physical conditions.

Yet, the phenomena globally identified under the label "turbulence" are so widespread and diverse that it is wise to expect, for many decades to come, ample room for studies and research. Moreover, the understanding and the knowledge, even complete, of a phenomenon do not automatically imply the ability to predict its quantitative details in a fast and economic way. To this purpose, one needs *models*, even approximate ones, that can give answers to practical problems with a reasonable amount of resources: this is why the development of turbulence models is still a field in continuous evolution and covers the bulk of the literature on turbulence.

Although the majority of the fluid dynamics problems encountered both in nature and in civil or industrial engineering regard turbulent flows, turbulence as a self-standing topic is usually absent from the syllabus of most university degrees (at least in Italy). Among other purposes, this work aims to fill this gap by providing the necessary theoretical and conceptual fundamentals and presenting a sufficiently complete panoramic of turbulence models, from the simplest to the most advanced ones including direct and large eddy simulation.

Since it is primarily addressed to engineers, the book avoids going to an excessive depth into the most abstruse and exoteric aspects of turbulence theory or into the vast field of experimental methods for turbulent flows, but rather focusses on problems of modelling and computation. Some information is provided, though, regarding the theory of dynamical systems and their bifurcations.

As a *caveat*, it must be stated that this work limits itself to the algebraic or differential formulation of mathematical models and does not treat the issue of their subsequent numerical resolution, for which several excellent books are available. Italian readers are addressed to *Fondamenti di Termofluidodinamica Computazionale* (Comini et al., eds, 2014). With few exceptions, at differential equation level, applying a turbulence model amounts to adding a certain number of scalar transport equations to the basic equations for laminar flow and does not need special numerical methods.

In regard to the contents of this book, Chap. 1 (*Introduction and governing equations*) provides a consistent basis for the treatment of turbulence and establishes a nomenclature. It is largely taken from the introductory lectures of the courses *Computational Thermofluid Dynamics* and *Models for Thermofluid Dynamics*, which I have held over several years at the University of Palermo for the Master's degrees in *Energy and Nuclear Engineering* and *Chemical Engineering*.

Chapters 2–5 are an expanded and updated version of my chapter on Turbulence in *Fondamenti di Termofluidodinamica Computazionale* (Comini et al., eds, 2014). In particular, Chap. 2 is an introduction on *decomposition, fluctuations* and *spectra*, while Chaps. 3–5 deal with *direct numerical simulation* (DNS), *large eddy simulation* (LES) and *Reynolds-averaged Navier–Stokes* (RANS) models, respectively.

Chapter 6 (*Turbulence in Free and Mixed Convection*) treats a class of problems that are often neglected in most textbooks on turbulence. It also attempts to throw a bridge towards the vast subject of atmospheric turbulence and meteorology, often made awkward to engineering students by flashy, but in the end only secondary, differences of approach and notation.

Chapters 7 and 8 are probably the most significant novelty of the present work and stem from keynote lectures I gave at two UIT (*Unione Italiana di Termofluidodinamica*) national heat transfer conferences, held in Brescia (Ciofalo and Di Liberto 2010) and in Bologna (Ciofalo et al. 2012).

Chapter 7 deals with *transient turbulence* (a condition scarcely discussed in most books on turbulence), in which the flow field is not stationary, even in a statistical sense. The conceptual problems arising from applying conventional Reynolds decomposition to these situations are discussed, and more appropriate *phase* or *ensemble* averages are defined. A specific model problem (reciprocating turbulent flow in a plane channel) is treated in detail.

Chapter 8 is dedicated to *transition to turbulence*. Following a brief introduction on bifurcations, the scenarios leading from steady laminar to chaotic (turbulent) conditions are discussed in some detail for specific configurations (toroidal pipe, serpentine pipe, spacer-filled channel) and intermediate steady, periodic and quasi-periodic regimes are identified.

An *Appendix* was included on the tensorial formulation of the main conservation and constitutive equations, mainly with the purpose of freeing the body of the text from too many mathematical definitions.

The present book can only scratch the surface of hydrodynamic turbulence, one of the most complex and fascinating fields of physical sciences. Books such as Tennekes and Lumley (1972), Hinze (1975), Lesieur (1990) and Pope (2000) provide precious complementary information. An updated state-of-the-art review of turbulence models is provided by Argyropoulos and Markatos (2015).

In addition, I warmly recommend whomever is interested in the study of turbulence spending some time watching real fluids in turbulent motion (of which nature, engineering and even home offer abundant examples); and to keep in mind that many familiar phenomena, such as the shape of the clouds that gained the attention of Hamlet and Polonius in *Hamlet* (Act III, Scene 2), challenge our current predictive capability.

Palermo, Italy   Michele Ciofalo

References

Argyropoulos CD, Markatos NC (2015) Recent advances on the numerical modelling of turbulent flows. Appl Math Mod 39(2):693–732

Ciofalo M, Di Liberto M (2010) Unsteady turbulence: phenomena and modelling. In: Pilotelli M, Beretta GP (eds) Proceedings of the 28th UIT heat transfer conference, Brescia, Italy, 20–23 June 2010. Snoopy publisher, Brescia, pp 15–28

Ciofalo M, Di Liberto M, Di Piazza I (2012) Steady, periodic, quasi-periodic and chaotic flow regimes in toroidal pipes. In: Lazzari S, Rossi di Schio E (eds) Proceedings of the 30th UIT heat transfer conference, Bologna, Italy, 25–27 June 2012. Esculapio, Bologna, pp 5–16

Comini G, Croce G, Nobile E (eds) (2014) Fondamenti di Termofluidodinamica Computazionale, 4th edn. Servizi Grafici Editoriali, Padova (in Italian)

Hinze JO (1975) Turbulence, 2nd edn. McGraw-Hill, New York

Lesieur M (1990) Turbulence in fluids. Kluwer Academic Publishers, Dordrecht

Pope SB (2000) Turbulent flows. Cambridge University Press, Cambridge, UK

Tennekes H, Lumley JL (1972) A first course in turbulence. MIT Press, Cambridge, MA

Acknowledgements

I would like to express my gratitude to the many colleagues, students and coworkers who have contributed, in different ways, to this work.

Michael W. Collins, in far 1985, made me aware for the first time of the fascinating notion that a numerical simulation can indeed reproduce the chaotic and pseudo-casual aspects of real-world turbulent physical systems, thus directing my entire subsequent research work towards these subjects. In the following years, until his passing away in 2014, Michael has been for me a constant source of inspiration and a dear friend.

Gianni Comini, already in 2001, persuaded me and the other authors to write a book on computational thermofluid dynamics, of which turbulence modelling was regarded as a crucial component, and in the following years has played an invaluable role of encouragement and coordination so that it kept being updated and enriched up to its 4th edition published in 2014.

Gian Paolo Beretta and Antonio Barletta, in inviting me to present my research work at the two UIT conferences mentioned above, stimulated me to put in order those ideas on transient turbulence and transition to turbulence that later became Chapters 7 and 8 of this book.

Alberto Brucato, Andrea Cipollina, Massimiliano Di Liberto, Ivan Di Piazza, Tadek Fodemski, Franco Grisafi, Luigi Gurreri, Tassos Karayiannis, Mariagiorgia La Cerva, Giorgio Micale, Jan Stasiek, Alessandro Tamburini and many others have shared with me through the years a warm interest in fluid dynamics and turbulence problems.

Acknowledgement

Contents

About the Author

Michele Ciofalo (b. 1951) is Professor of Nuclear Engineering at the University of Palermo, Italy, where he lectures in *Thermal-Hydraulics, Models for Thermofluid Dynamics* and *Computational Thermofluid Dynamics*. His most recent research interests include fluid flow and heat or mass transfer in membrane processes such as membrane distillation, direct and reverse electrodialysis and acid–base energy storage devices.

He is the author of about 100 journal papers and 100 conference papers, besides several books or book sections including *"Nanoscale Fluid Dynamics in Physiological Flows—A Review Study"* (with Michael W. Collins and Tom R. Hennessy, published by WIT Press in 1999) and the chapter on *Turbulence* in *"Fondamenti di Termofluidodinamica Computazionale"* (published by SGE in 2014).

Symbols and Abbreviations

A	Cross-sectional area
A^+	Van Driest constant
$\mathbf{a}; a_i$	Acceleration
$a(n), a(f)$	Fourier transforms of $\varphi(x), \varphi(t)$
$a_1, \beta^*, \beta_i, c_{ki}, c_{\omega i}\ (i = 1, 2)$	Constants in the SST model
$\mathbf{B}; B_i$	Magnetic induction field
$\mathbb{B}; B_{ij}$	Auxiliary tensor in differential stress models
C, C_D, C_E	Constants in nonlinear $k-\varepsilon$ model
C_f	Fanning friction coefficient
C_S, C_{ij}	Cross terms
$C_1, C_2, C_3, C_\mu, \sigma_k, \sigma_\varepsilon$	Constants in the k-ε model
$C_2^0, \beta_{RNG}, \eta_0$	Additional constants in the RNG k-ε model
$C_{\Phi 1}, C_{1w}, C_{2w}, \gamma_t, c_t$	Constants in differential stress (DS) models
$C_{\varepsilon 1}, C_{\varepsilon 2}, C_\varepsilon$	Constants in the equation for ε in DS models
C_θ	Constant in Daly–Harlow GGDH model
$CD_{k\omega}$	Term in the *SST* model
c	Curvature radius of a curved duct
c_p	Specific heat at constant pressure
c_S	Constant in the Smagorinsky model
D	Kinematic diffusivity, Γ/ρ
D_{eq}	Equivalent (hydraulic) diameter
D_{ij}	Oldroyd derivatives of strain rates
$D_\varepsilon, E_\varepsilon, f_1, f_2$	Terms in low-Reynolds number k-ε models
d	Pipe diameter
De	Dean number, Re $\sqrt{\delta}$
E	Constant in universal near-wall profiles
F	Dimensionless frequency (Strouhal number)
F_1	Blending function in the *SST* model
f	Frequency
f_D	Darcy friction coefficient
f_K	Kolmogorov frequency of dissipative eddies

f_μ	Near-wall damping factor
f_0	Amplitude of the forcing term in pulsatile flow
f_1	Shear force associated with one eddy
$\mathbf{f}_S; f_{S,i}$	Surface force per unit surface area
$\mathbf{f}_V; f_{V,i}$	Volume force per unit volume
$G(\mathbf{x}, \mathbf{y})$	Filtering function
G_k	Buoyancy production of ρk in eddy viscosity models
G_{ij}	Reynolds stress buoyancy production terms in DS models
$\mathbf{g}; g_i$	Gravity acceleration
H	Height of a plane channel
h	Local heat transfer coefficient, $q_w/(T_w-T_b)$
J	Enthalpy per unit mass
$\mathbf{j}; j_i$	Diffusive flux of a scalar
K	Clauser constant, ~0.0168
K_x, K_y, K_z	Dimensions of computational box normalized by δ
K_t	Total simulated time in LETOT's δ/u_τ
k	Turbulent kinetic energy per unit mass, $1/2\langle u_i' u_i'\rangle$ (RANS)
k_{sg}	Subgrid kinetic energy per unit mass, $1/2\langle u_i' u_i'\rangle$ (LES)
L	Largest turbulence length scale
L_{MO}	Monin–Obukhov length
L_x, L_y, L_z	Dimensions of computational box
L_S, L_{ij}	Leonard terms
$L_{ij}^*, T_{ij}^*, M_{ij}^*$	Auxiliary stresses in the dynamic subgrid model for LES
Q_i^*, N_i^*	Auxiliary scalar fluxes in dynamic subgrid model for LES
l	Eddy length scale; Prandtl mixing length
l_S	Stokes length, $(2\nu/\omega)^{1/2}$
l_ω	Length of a vorticity tube
m	Mass
$\dot{m}$	Mass flow rate
N	Number of realizations/periods; dimensions of phase space
N_{FLOP}	Number of floating point operations
N_p	Number of grid points
N_t	Number of time steps
n	Wavenumber
$\mathbf{n}; n_i$	Versor normal to a surface and pointing outward
Nu	Local Nusselt number, hD_{eq}/λ
P	Pitch
P_J	Jayatilleke function in universal near-wall scalar profiles
P_k	Shear production of ρk in eddy viscosity models

P_{ij}	Reynolds stress shear production terms in DS models
$\mathbb{P}; \wp_{ij}$	Stress tensor
p	Thermodynamic pressure
$\tilde{p}$	Non-hydrostatic pressure component, $p-\rho a_k x_k$
p_{sg}	Subgrid pressure, $(2/3)\rho k_{sg}$ (LES)
p_t	Turbulent pressure, $(2/3)\rho k$ (RANS)
p^*	Modified pressure, $p+2/3\mu\nabla\cdot\mathbf{u}$
p^{**}	Total pressure, $\langle p^*\rangle+(2/3)\rho k_{sg}$ (LES); $\langle p^*\rangle+(2/3)\rho k$ (RANS)
Q_1	Thermal power transferred by a single eddy
$\mathbf{q}; q_i$	Heat flux
q_w	Wall heat flux
q'''	Volumetric power density
R	Pipe radius
R'	Gas constant for a specific gas
R_{ij}	Subgrid stresses (LES); Reynolds stresses (RANS)
R_k, R_t	Auxiliary Reynolds numbers in low-Re k-ε models
$R_{\varphi\varphi}(x), R_{\varphi\varphi}(t)$	Auto-correlation functions of $_\varphi(x)$, $_\varphi(t)$
r	Radial coordinate
r_{ext}	Thermal resistance in Robin thermal boundary conditions
Re	Bulk Reynolds number, UD_{eq}/ν
Re$_{cr}$	Critical Reynolds number for regime transition
Re$_{peak}$	Peak velocity Reynolds number, $\bar{u}_{peak}D_{eq}/\nu$
Re$_S$	Stokes length Reynolds number, $\bar{u}_{peak}l_S/\nu$
Re$_\tau$	Friction velocity Reynolds number, $u_\tau\delta/\nu$
Ri	Richardson number, $-G_k/P_k$
S	Surface area
$\mathbb{S}; S_{ij}$	Strain rate tensor, $1/2(\partial u_i/\partial x_j+\partial u_j/\partial x_i)$
S_{max}	Computer peak speed in FLOPS
S_φ	Volumetric source of a scalar φ
s	Axial coordinate
Sc	Schmidt number, $\mu/\Gamma = \nu/D$
T	Absolute temperature
T_b	Bulk temperature
T_w	Wall temperature
t	Time
t_{avg}	Averaging time
t_{per}	Periodic time
t_{TOT}	Total simulated time
t_{trans}	Time scale of a transient
tt	Throughput time
U	Mean velocity (averaged over time and cross section)
$\mathbf{u}; u_i; u, v, w$	Velocity vector and its Cartesian components
u_{abs}	Absolute velocity in oscillatory flow

u_e	Velocity at the edge of a boundary layer
u_{peak}	Peak value of velocity u with respect to time or phase
u_s	Velocity component along the axial direction
u_w	Velocity of the wall
u_τ	Friction velocity, $(\tau_w/\rho)^{1/2}$
u^*	Conventional friction velocity, $C_\mu^{1/4} k_P^{1/2}$
V	Volume
v	Eddy peripheral velocity; rms fluctuating velocity
W	Spanwise extent
$\mathbf{x}; x_i; x, y, z$	Position vector and Cartesian coordinates
x_w	Displacement of the wall in oscillatory flow
x^*	Location of reattachment point
X	Generic extensive scalar quantity
y	Distance from the nearest wall
y_v, y_T	Thickness of viscous and conductive sublayers
$Z_{\varphi\varphi}(n), Z_{\varphi\varphi}(f)$	Spectral density of $\varphi(x), \varphi(t)$

Greek Symbols

α	Dimensionless acceleration, ad^3/ν^2
α_T	Kinematic thermal diffusivity, $\lambda/(\rho c_p)$
α_W	Womersley number, $\delta\,(\omega/\nu)^{1/2}$
$\beta, \beta^*, \sigma_k, \sigma_\omega$	Constants in the $k-\omega$ model
β_E	Enstrophy dissipation
β_T	Cubic dilatation coefficient
$\beta_W, \varphi_1, \varphi_2$	Auxiliary terms in laminar oscillatory flow solutions
Γ	Diffusivity
Γ_a	Dry adiabatic temperature gradient, g/c_p
Γ_{sg}	Subgrid diffusivity in LES
Γ_t	Turbulent (eddy) diffusivity in RANS models
γ	Ratio between specific heats c_p and c_v
γ_I	Klebanoff intermittence factor
Δ	Width of a spatial filter in LES
$\Delta x_i; \Delta x, \Delta y, \Delta z$	Dimensions of cells in a computational grid
Δt	Time step
δ	Half-height of a plane channel, $H/2$; curvature, R/c
$\delta_{BL}, \delta_{BL}^*$	Kinematic and momentum thicknesses of a boundary layer
δ_{ij}	Kronecker's delta
ε	Dissipation of turbulent kinetic energy per unit mass
ε^*	Modified dissipation in low-Re models, $\varepsilon - D_\varepsilon$
ζ	Ratio of test to grid filters in subgrid dynamic model for LES
η	Ratio S/ω in RNG k-ε model (RANS)
η_B	Batchelor length scale, $\eta_K/Sc^{1/2}(Sc>1)$

η_K	Kolmogorov length scale of dissipative eddies, $(\nu^3/\varepsilon)^{1/4}$
η_{OC}	Obukhov–Corrsin length scale, $\eta_K/Sc^{3/4}$ (Sc<1)
ϑ	Flow attack angle in spacer-filled channels
θ	Potential temperature
κ	Von Karman constant, ~0.42
Λ	Wavelength
λ	Thermal conductivity; control parameter
μ	Viscosity
μ_{sg}	Subgrid viscosity in LES
μ_t	Turbulent viscosity in RANS models
ν	Kinematic viscosity, μ/ρ
ξ	Fraction of time spent in turbulent state by oscillatory flow
ρ	Density
$\Sigma(n)$	Rate of spectral energy transfer through wavenumber n
σ	Prandtl number, $\nu/\alpha = c_p\mu/\lambda$
σ_{sg}	Subgrid Prandtl number in LES, μ_{sg}/Γ_{sg}
σ_t	Turbulent Prandtl number in RANS models, μ_t/Γ_t
τ; τ_{ij}	Traceless part of the stress tensor
τ_t	Turbulent shear stress
τ_w	Wall shear stress
Φ	Dissipation function
$(\Phi_{ij})_1, (\Phi_{ij})_2, (\Phi_{ij})_3$	Redistribution terms in Reynolds stress models
ϕ	Azimuthal coordinate
φ	Generic quantity (in particular, intensive scalar)
χ	Computational efficiency (ratio of actual to peak speed)
$\mathbf{\Omega}$; Ω_{ij}	Vorticity tensor
$\boldsymbol{\omega}$; ω_i	Vorticity vector (curl of velocity)
ω	Angular velocity; eddy frequency ε/k (RANS); pulsation

Superscripts

+	Expressed in wall units
*	Reattachment point; auxiliary or dimensionless variable
I, II	Oscillatory modes in periodic and quasi-periodic flows
(k)	Generic realization
rms	Root mean square value
$\dot{x}$	First derivative of x with respect to time
$\ddot{x}$	Second derivative with respect to time

Subscripts

i, j, k	Coordinate directions
peak	Maximum (with respect to time or phase angle)
s, r, ϕ	Axial, radial and azimuthal directions (curved pipes)
sg	Subgrid
T	Thermal
t	Turbulent
x, y, z	Cartesian directions
0	Reference value

Averages and Statistics

$\langle \varphi \rangle$	Mean or filtered value of φ over time, space or phase
φ'	Fluctuating or residual value of φ, $\varphi - \langle \varphi \rangle$
$\bar{\varphi}$	Average of φ over a cross section

Abbreviations and Acronyms

ASM	Algebraic stress model
DNS	Direct numerical simulation
DOF	Degrees of freedom
DSM	Differential stress model
FLOP	Floating-point operations
FLOPS	Floating-point operations per second
LES	Large eddy simulation
LETOT	Large eddy turnover time, δ/u_τ
LHS	Left-hand side
RAM	Random access memory
RANS	Reynolds-averaged Navier–Stokes
RHS	Right-hand side
RNG	Re-normalization group
SST	Shear stress transport

Multiples:
k(ilo) = 10^3; M(ega) = 10^6; G(iga) = 10^9; T(era) = 10^{12}; P(eta) = 10^{15}; E(xa) = 10^{18}

Chapter 1
Introduction and Governing Equations

Rien n'est aussi pratique qu'une bonne théorie
Kurt Levin

Abstract The differential equations governing the transport of mass, momentum and energy (or other scalars) in a flowing fluid are derived from the corresponding dynamic balances, written for a macroscopic volume V bounded by a surface S and embedded in a velocity field u(x, y, z, t). The special case of constant-property fluids and the Boussinesq approximation for buoyancy are also presented.

Keywords Continuity equation · Navier–Stokes equations · Energy equation · Boussinesq approximation · Newtonian fluid · Scalar transport

1.1 Eulerian and Lagrangian Approaches

For the sake of completeness, two alternative but equivalent approaches will be considered. In the first (Eulerian approach, Fig. 1.1a), the volume V is fixed with respect to an inertial reference frame ($Oxyz$) but is an open, variable-mass volume whose boundary can be freely crossed by the fluid. In the second (Lagrangian approach, Fig. 1.1b), the volume V' is closed (and thus contains a constant mass of fluid), but is dragged by the local velocity field. In both cases the versor normal to the surface S and pointing outwards will be denoted by **n**. The Cartesian tensor notation will be preferentially used (see Appendix).

1.1.1 Continuity

The continuity equation is more easily derived considering the open control volume V of Fig. 1.1a. The mass balance in V can be expressed as:

© The Author(s), under exclusive license to Springer Nature Switzerland AG 2022
M. Ciofalo, *Thermofluid Dynamics of Turbulent Flows*, UNIPA Springer Series,
https://doi.org/10.1007/978-3-030-81078-8_1

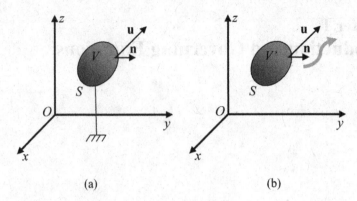

Fig. 1.1 Control volumes for the derivation of mass, momentum and energy balances. **a** Control volume open and fixed with respect to the reference frame. **b** Control volume closed and dragged by the fluid velocity field

$$\frac{\partial m}{\partial t} = -\int_S \rho u_j n_j dS \tag{1.1}$$

in which m is the mass instantaneously contained in the volume, and the integral represents the mass flow exiting the volume through the surface S.

The mass m can be written as

$$m = \int_V \rho dV \tag{1.2}$$

so that

$$\frac{\partial m}{\partial t} = \frac{\partial}{\partial t} \int_V \rho dV = \int_V \frac{\partial \rho}{\partial t} dV \tag{1.3}$$

in which use was made of the possibility of inverting the order of integration and derivation.

According to the Gauss theorem, the surface integral of the flux of vector ρu_j at the right hand side of (1.1) is equal to the volume integral of its divergence:

$$\int_S \rho u_j n_j dS = \int_V \frac{\partial \rho u_j}{\partial x_j} dV \tag{1.4}$$

By substituting Eqs. (1.3) and (1.4) into Eq. (1,1), the mass balance becomes

$$\int_V \frac{\partial \rho}{\partial t} dV = -\int_V \frac{\partial \rho u_j}{\partial x_j} dV \tag{1.5}$$

which can also be written

$$\int_V \left(\frac{\partial \rho}{\partial t} + \frac{\partial \rho u_j}{\partial x_j} \right) dV = 0 \tag{1.6}$$

Since this last equation must hold for any arbitrary choice of volume V, the following must be true:

$$\frac{\partial \rho}{\partial t} + \frac{\partial \rho u_j}{\partial x_j} = 0 \tag{1.7}$$

Equation (1.7) is the most general form of the continuity equation, provided the existence of mass sources is barred. For a steady-state problem ($\partial/\partial t = 0$) it simplifies to

$$\frac{\partial \rho u_j}{\partial x_j} = 0 \tag{1.8}$$

while, if the density ρ is a constant, it simplifies to

$$\frac{\partial u_j}{\partial x_j} = 0 \tag{1.9}$$

indipendent of the flow being stationary or not. This last equation can be written in vector notation as $\nabla \mathbf{u} = 0$ and states that, in a constant density fluid, the velocity field is always divergence-free (solenoidal).

Note that constant density fluids are often improperly called *incompressible*. In rigorous terms a fluid is incompressible if

$$\left(\frac{\partial \rho}{\partial p} \right)_{T=\text{const.}} = 0 \tag{1.10}$$

i.e., if its density does not change with pressure at constant temperature. A fluid being incompressible does not exclude that its density may significantly vary from point to point and with time, e.g. as the effect of temperature changes; in this case, the velocity field will *not* be divergence-free.

1.1.2 Momentum

In view of their importance and complexity, the momentum trnsport equations will be derived here following both the approaches mentioned above.

(a) *Conservation form: derivation from the momentum balance for an open volume fixed in the reference frame (Eulerian approach).*

For the volume V in Fig. 1.1a (which is fixed in the laboratory reference frame, and whose walls can freely be crossed by the fluid), the momentum balance in the generic direction i can be written:

$$\frac{\partial}{\partial t} \int_V \rho u_i dV = - \int_S \rho u_j u_i n_j dS + \int_S f_{S,i} dS + \int_S f_{V,i} dV \tag{1.11}$$

which expresses a dynamic balance between the ith momentum component of the masses in V, its net flow *into* V through the boundary surface S, and the ith component of the resultant of forces acting on V, including both surface and volume ones. The explicit minus sign in the first term of the right hand side accounts for the fact that, by general convention, n_i is the versor normal to S and directed *outwards*.

The volume force can be expressed as $f_{V,i} = \rho a_i$, in which a_i is an acceleration (usually, gravity). As discussed in the Appendix, Section A.4 (Stresses and constitutive equations), surface forces per unit area $f_{s,i}$ are related to the stress tensor $\wp_{ij}$ by $f_{s,i} = \wp_{ij} n_j$. Therefore, Eq. (1.11) can be written:

$$\frac{\partial}{\partial t} \int_V \rho u_{i} dV = - \int_S \rho u_j u_i n_j dS + \int_S \wp_{ij} n_j dS + \int_V \rho a_i dV \tag{1.12}$$

By applying the Gauss theorem to surface integrals, this becomes:

$$\frac{\partial}{\partial t} \int_V \rho u_i dV = - \int_V \frac{\partial}{\partial x_j} \rho u_j u_i dV + \int_V \frac{\partial}{\partial x_j} \wp_{ij} dV + \int_V \rho a_i dV \tag{1.13}$$

Now, by following an argument similar to that applied to the continuity equation, Eqs. (1.5)–(1.7), and moving the first term at the RHS of Eq. (1.13) to the LHS, the following differential equation is obtained:

$$\frac{\partial \rho u_i}{\partial t} + \frac{\partial \rho u_j u_i}{\partial x_j} = \frac{\partial \wp_{i,j}}{\partial x_j} + \rho a_i \tag{1.14}$$

The stress tensor can always be decomposed into an isotropic pressure term and a traceless part, i.e. $\wp_{ij} = -\delta_{ij} p + \delta_{ij}$ (δ_{ij} being Kronecker's delta). By substituting this last expression for $\wp_{ij}$ into Eq. (1.14), this can be reduced with simple manipulations to

$$\frac{\partial \rho u_i}{\partial t} + \frac{\partial \rho u_j u_i}{\partial x_j} = -\frac{\partial p}{\partial x_i} + \frac{\partial \tau_{ij}}{\partial x_j} + \rho a_i \qquad (1.15)$$

This is the most general form of the ith momentum equation and applies to all fluids, including compressible and non-Newtonian ones.

If the fluid is Newtonian, as discussed in the Appendix, Section A.4, the constitutive equation $\tau_{ij} = 2\mu(S_{ij} - \frac{2}{3}\delta_{ij} S_{kk})$ applies, in which μ is the fluid's viscosity and S_{ij} is the strain rate tensor. By substituting this last expression for τ_{ij} into Eq. (1.15) one obtains:

$$\frac{\partial \rho u_i}{\partial t} + \frac{\partial \rho u_j u_i}{\partial x_j} = -\frac{\partial}{\partial x_i}\left(p + \frac{2}{3}\mu S_{kk}\right) + \frac{\partial}{\partial x_j}(2\mu S_{ij}) + \rho a_i \qquad (1.16)$$

By defining a modified pressure $p^* = p + \frac{2}{3}\mu S_{kk} = p + \frac{2}{3}\mu \nabla \mathbf{u}$ and writing the strain rate tensor S_{ij} explicitly (see Appendix, Section A.3), Eq. (1.16) becomes

$$\frac{\partial \rho u_i}{\partial t} + \frac{\partial \rho u_j u_i}{\partial x_j} = -\frac{\partial p^*}{\partial x_i} + \frac{\partial}{\partial x_j}\left[\mu\left(\frac{\partial u_i}{\partial x_j} + \frac{\partial u_j}{\partial x_i}\right)\right] + \rho a_i \qquad (1.17)$$

This is the ith *Navier–Stokes equation*, or momentum equation for a Newtonian fluid, written in the so-called *conservation form*. Of course, there are three such equations which, together with the *continuity Eq.* (1.7) and (if required, i.e. in variable density fluids) with an *equation of state* of the form $\rho = f(p)$, balance the three velocity components u_i ($i = 1, 2, 3$), pressure p and density ρ.

(b) *Non-conservation form: derivation from Newton's second law for a closed volume dragged by the fluid (Lagrangian approach).*

Consider now a volume V' which, unlike volume V of Fig. 1.1a, is closed (and thus contains a constant mass of fluid), but is dragged by the fluid's velocity field (Fig. 1.1b).

Newton's second law ($\mathbf{f} = m\mathbf{a} = m d\mathbf{u}/dt$), written for the generic direction i and integrated over V', yields:

$$\int_{V'} \rho dV \frac{du_i}{dt} = \int_{S} f_{S,i} dS + \int_{V'} \rho a_i dV \qquad (1.18)$$

By explicitly writing the material derivative as

$$\frac{du_i}{dt} = \frac{\partial u_i}{\partial t} + u_j \frac{\partial u_i}{\partial x_j} \qquad (1.19)$$

Equation (1.18) becomes

$$\int_{V'} \rho\left(\frac{\partial u_i}{\partial t} + u_j \frac{\partial u_i}{\partial x_j}\right) dV = \int_S f_{S,i} dS + \int_{V'} \rho a_i dV \qquad (1.20)$$

Note that, by contrast with Eq. (1.11), in Eq. (1.20) the term expressing the advective flow of momentum through the volume's surface S is missing from the RHS because S is now impermeable to the fluid, but velocities appear anyway at the LHS as components of the material derivative.

Proceeding now as in the previous case (i.e. applying the Gauss theorem, writing surface forces in terms of the stress tensor and moving from the integral form to the differential one), the following equation is obtained:

$$\rho\frac{\partial u_i}{\partial t} + \rho u_j \frac{\partial u_i}{\partial x_j} = -\frac{\partial p}{\partial x_i} + \frac{\partial \tau_{ij}}{\partial x_j} + \rho a_i \qquad (1.21)$$

which replaces Eq. (1.15) as the most general form of the ith momentum equation for arbitrary fluids.

In the case of Newtonian fluids, treating the RHS of Eq. (1.21) exactly as in the previous approach, the so-called *non conservation form* of the generic Navier–Stokes equation is obtained:

$$\rho\frac{\partial u_i}{\partial t} + \rho u_j \frac{\partial u_i}{\partial x_j} = -\frac{\partial p^*}{\partial x_i} + \frac{\partial}{\partial x_j}\mu\left(\frac{\partial u_i}{\partial x_j} + \frac{\partial u_j}{\partial x_i}\right) + \rho a_i \qquad (1.22)$$

Equations (1.15) and (1.21), or (1.17) and (1.22), differ only in their left hand sides. The two forms are actually equivalent: in fact, starting for example from the conservation form and developing the derivatives of products, one has:

$$\frac{\partial \rho u_i}{\partial t} + \frac{\partial \rho u_j u_i}{\partial x_j} = \rho\frac{\partial u_i}{\partial t} + u_i\frac{\partial \rho}{\partial t} + \rho u_j\frac{\partial u_i}{\partial x_j} + u_i\frac{\partial \rho u_j}{\partial x_j}$$

$$= \rho\frac{\partial u_i}{\partial t} + \rho u_j\frac{\partial u_i}{\partial x_j} + u_i\left(\frac{\partial \rho}{\partial t} + \frac{\partial \rho u_j}{\partial x_j}\right) = \rho\frac{\partial u_i}{\partial t} + \rho u_j\frac{\partial u_i}{\partial x_j}$$

$$(1.23)$$

since the term in parentheses vanishes because of the continuity Eq. (1.7).

The conservation form is often used as the starting point for the finite volume formulation of the primitive equations, whereas the non conservation form is used at the starting point for the stream function-vorticity (ψ–ω) formulation.

1.1.3 Scalar Transport

Consider an *extensive* scalar quantity X and let its concentration per unit mass (an *intensive* scalar) be φ. For example, if X is enthalpy in J, φ will be specific enthalpy in J/kg; if X is the amount of a solute in kg, φ will be its concentration in kg/kg.

The scalar φ is called *passive* if the flow equations (continuity and momentum) do not depend on φ; contrariwise, it is called *participating*. If the scalar is passive, the coupling between flow equations and φ-transport equation is *one-way*: velocities are necessary to evaluate advection terms and thus the distribution of φ, but there is no feedback from φ to the flow field.

The main mechanisms that may make a scalar φ a participating one are:

- The fluid's physical properties depend on φ (for example, the dependence of density on temperature may give rise to buoyancy forces which cause or affect the fluid's motion; the dependence of viscosity on temperature may modify the structure of near-wall layers and affect friction and heat transfer coefficients; etc.).
- The presence of φ causes or modifies volume forces (for example, electrophoretic or magnetohydrodynamic forces that may depend on the concentration of an electrolyte).

Also the scalar transport equation, like the momentum ones, can be derived following two different approaches.

(a) *Conservation form: derivation from the balance of scalar φ in an open volume fixed in the reference frame (Eulerian approach).*

With reference to the same volume V in Fig. 1.1a (fixed in the laboratory reference frame, and with walls freely crossed by the fluid), let **j** be the diffusive flux of φ (which is always present also in mainly advective problems!) and S_φ its volumetric source (if any). The balance of the quantity X in V can be written:

$$\frac{\partial}{\partial t} \int_V \rho\varphi dV = -\int_S \rho\varphi u_j n_j dS - \int_S j_j n_j dS + \int_V S_\varphi dV \qquad (1.24)$$

As for momentum, the explicit minus sign in the first two terms at the RHS accounts for the fact that n_j is the versor normal to S and pointing *outwards*.

By applying the Gauss theorem to surface integrals, Eq. (1.24) becomes:

$$\frac{\partial}{\partial t} \int_V \rho\varphi dV = -\int_V \frac{\partial}{\partial x_j}\rho\varphi u_j dV - \int_V \frac{\partial}{\partial x_j} j_j dV + \int_V S_\varphi dV \qquad (1.25)$$

By assuming that the diffusive flux **j** can be expressed by Fick's law as $j_j = -\Gamma \partial\varphi/\partial x_j$, in which Γ is the scalar's diffusivity (see Appendix), Eq. (1.25) becomes:

$$\frac{\partial}{\partial t}\int_V \rho\varphi dV = -\int_V \frac{\partial}{\partial x_j}\rho\varphi u_j dV - \int_V \frac{\partial}{\partial x_j}\left(-\Gamma\frac{\partial\varphi}{\partial x_j}\right)dV + \int_V S_\varphi dV \quad (1.26)$$

Now, by steps similar to those followed for deriving the continuity equation, Eqs. (1.5)–(1.7), and the momentum equation, Eqs. (1.13)–(1.15), the integral equation (1.26) gives rise to the differential equation:

$$\frac{\partial\rho\varphi}{\partial t} + \frac{\partial\rho u_j\varphi}{\partial x_j} = \frac{\partial}{\partial x_j}\left(\Gamma\frac{\partial\varphi}{\partial x_j}\right) + S_\varphi \quad (1.27)$$

which is the *conservation form* of the scalar transport equation for φ.

The energy equation can be regarded as the special form that Eq. (1.27) takes when $\varphi = J = c_p T$ (specific enthalpy), $\Gamma = \lambda/c_p$ (λ being the fluid's thermal conductivity) and $S_\varphi = q'''$ (volumetric power density). In conservation form, this is:

$$\frac{\partial\rho c_p T}{\partial t} + \frac{\partial\rho c_p u_j T}{\partial x_j} = \frac{\partial}{\partial x_j}\left(\lambda\frac{\partial T}{\partial x_j}\right) + q''' \quad (1.28)$$

(b)   *Non-conservation form: derivation from the balance of a scalar φ for a closed volume dragged by the fluid (Lagrangian approach).*

By proceeding as for momentum, consider now volume V' in Fig. 1.1b which, unlike volume V in Fig. 1.1a, has impermeable walls and thus contains a constant mass of fluid, but is dragged by the local velocity field. The X balance is now:

$$\int_{V'} \rho\frac{d\varphi}{dt}dV = -\int_S j_j n_j dS + \int_{V'} S_\varphi dV \quad (1.29)$$

and differs from Eq. (1.24) for the presence of the material derivative *in lieu* of the partial one and for the absence of advective fluxes through the bounding surface S. Diffusive fluxes are still present because the surface S is assumed to be impermeable to the fluid but permeable to diffusion.

By writing the material derivative explicitly according to Eq. (1.19), as for momentum, Eq. (1.29) becomes:

$$\int_{V'} \rho\left(\frac{\partial\varphi}{\partial t} + u_j\frac{\partial\varphi}{\partial x_j}\right)dV = -\int_S j_j n_j dS + \int_{V'} S_\varphi dV \quad (1.30)$$

Proceeding now as in the previous approach (i.e. applying the Gauss theorem, expressing diffusive fluxes by Fick's law and switching from the integral form to the differential one) one obtains:

$$\rho \frac{\partial \varphi}{\partial t} + \rho u_j \frac{\partial \varphi}{\partial x_j} = \frac{\partial}{\partial x_j}\left(\Gamma \frac{\partial \varphi}{\partial x_j}\right) + S_\varphi \tag{1.31}$$

Equation (1.31) is the so called *non conservation* form of a scalar transport equation. As in the case of the momentum equations, the two forms are equivalent; in fact, by expliciting the derivatives of products at the LHS, one has:

$$\frac{\partial \rho \varphi}{\partial t} + \frac{\partial \rho u_j \varphi}{\partial x_j} = \rho \frac{\partial \varphi}{\partial t} + \varphi \frac{\partial \rho}{\partial t} + \rho u_j \frac{\partial \varphi}{\partial x_j} + \varphi \frac{\partial \rho u_j}{\partial x_j}$$

$$= \rho \frac{\partial \varphi}{\partial t} + \rho u_j \frac{\partial \varphi}{\partial x_j} + \varphi \left(\frac{\partial \rho}{\partial t} + \frac{\partial \rho u_j}{\partial x_j}\right) = \rho \frac{\partial \varphi}{\partial t} + \rho u_j \frac{\partial \varphi}{\partial x_j} \tag{1.32}$$

since the term in parentheses vanishes due to the continuity Eq. (1.7).

In non-conservation form, the energy Eq. (1.28) will be written

$$\rho \frac{\partial c_p T}{\partial t} + \rho u_j \frac{\partial c_p T}{\partial x_j} = \frac{\partial}{\partial x_j}\left(\lambda \frac{\partial T}{\partial x_j}\right) + q''' \tag{1.33}$$

1.1.4 Governing Equations for Constant-Property Fluids

In the case of constant physical properties, the governing Eqs. (1.7) (continuity), (1.17) or (1.22) (Navier–Stokes, in conservation or non-conservation form, respectively) and (1.27) or (1.31) (scalar transport, in conservation or non-conservation form, respectively) can be somewhat simplified.

In regard to the continuity equation, it was already stated in Sect. 1.1 that, for $\rho = $ constant, it takes the form (1.9) ($\nabla \mathbf{u} = 0$), indipendent of the flow field being stationary or not.

In regard to the generic (*i*th) Navier–Stokes equation—written, for example, in the conservation form (1.17)—the assumption $\rho = $ constant, which implies $\nabla \mathbf{u} = 0$, causes the modified pressure $p^* = p + {}^2\!/_3 \mu \nabla \mathbf{u}$ to coincide with the thermodynamic pressure p. If also $\mu = $ constant, the viscous term can be simplified by extracting μ from the derivative and switching the order of derivation in the second term:

$$\frac{\partial}{\partial x_j} \mu \left(\frac{\partial u_i}{\partial x_j} + \frac{\partial u_j}{\partial x_i}\right) = \mu \frac{\partial}{\partial x_j}\frac{\partial u_i}{\partial x_j} + \mu \frac{\partial}{\partial x_i}\frac{\partial u_j}{\partial x_j} = \mu \frac{\partial^2 u_i}{\partial x_j^2} \tag{1.34}$$

Therefore, dividing all terms by ρ and defining the *kinematic viscosity* $\nu = \mu/\rho$, Eq. (1.17) becomes, in conservation form:

$$\frac{\partial u_i}{\partial t} + \frac{\partial u_i u_j}{\partial x_j} = -\frac{1}{\rho}\frac{\partial p}{\partial x_i} + v\frac{\partial^2 u_i}{\partial x_j^2} + a_i \tag{1.35}$$

The corresponding non-conservation form is:

$$\frac{\partial u_i}{\partial t} + u_j\frac{\partial u_i}{\partial x_j} = -\frac{1}{\rho}\frac{\partial p}{\partial x_i} + v\frac{\partial^2 u_i}{\partial x_j^2} + a_i \tag{1.36}$$

Similar simplifications are obtained in the scalar transport Eqs. (1.27) (conservation form) o (1.33) (non-conservation form) if ρ and Γ are constant. The resulting transport equations may be written in conservation form as

$$\frac{\partial \varphi}{\partial t} + \frac{\partial u_j \varphi}{\partial x_j} = \frac{\partial}{\partial x_j}\left(D\frac{\partial \varphi}{\partial x_j}\right) + \frac{S_\varphi}{\rho} \tag{1.37}$$

and, in non-conservation form, as

$$\frac{\partial \varphi}{\partial t} + u_j\frac{\partial \varphi}{\partial x_j} = \frac{\partial}{\partial x_j}\left(D\frac{\partial \varphi}{\partial x_j}\right) + \frac{S_\varphi}{\rho} \tag{1.38}$$

where $D = \Gamma/\rho$ is the kinematic diffusivity of the scalar φ (units m²/s).

In particular, for constant ρ, c_p and λ the enthalpy transport equation, or energy Eq. (1.28) becomes, in conservation form

$$\frac{\partial T}{\partial t} + \frac{\partial u_j T}{\partial x_j} = \alpha_T\frac{\partial^2 T}{\partial x_j^2} + \frac{q'''}{\rho c_p} \tag{1.39}$$

and, in non-conservation form

$$\frac{\partial T}{\partial t} + u_j\frac{\partial T}{\partial x_j} = \alpha_T\frac{\partial^2 T}{\partial x_j^2} + \frac{q'''}{\rho c_p} \tag{1.40}$$

in which $\alpha_T = \lambda/(\rho c_p)$ is the kinematic thermal diffusivity (units m²/s).

1.1.5 Boussinesq Approximation for Buoyancy

Buoyancy phenomena, responsible for free convection, occur in the simultaneous presence of a non-uniform density ρ and of a volume force **a** (per unit mass) such that $\mathbf{a} \cdot \nabla \rho \neq 0$.

In particular, thermal buoyancy occurs in a dilatable fluid, i.e. in a fluid whose density depends on temperature according to a certain law $\rho = \rho(T)$, provided a vertical temperature stratification is present.

Since buoyancy phenomena are related to changes in density and temperature, their complete simulation requires the use of the general form of the Navier–Stokes equations for variable-property fluids, i.e. Equations (1.17) or (1.22), coupled with the energy Eq. (1.28) or (1.33) and with the general continuity Eq. (1.7).

However, buoyancy effects can be predicted to a good approximation even using the constant-property form of the governing equations, provided suitable source terms are adopted. This approach is known as *Boussinesq approximation* for buoyancy.

Suppose density to be a linear function of temperature:

$$\rho = \rho_0[1 - \beta_T(T - T_0)] \tag{1.41}$$

in which ρ_0 is the density at a reference temperature T_0 and $\beta_T = -(1/\rho_0)\partial\rho/\partial T$ is the fluid's cubic dilatation coefficient. Using, for example, conservation forms, start from Eq. (1.17) and substitute Eq. (1.41) for ρ only in the term representing the volume force, while assuming $\rho = $ constant in all other terms. With some manipulation, the following form of the generic Navier–Stokes equation is obtained:

$$\frac{\partial u_i}{\partial t} + \frac{\partial u_j u_i}{\partial x_j} = -\frac{1}{\rho}\frac{\partial \tilde{p}}{\partial x_i} + \nu\frac{\partial^2 u_i}{\partial x_j^2} - a_i\beta(T - T_0) \tag{1.42}$$

in which $\tilde{p} = p - \rho a_k x_k$ is pressure epurated from the hydrostatic term $\rho a_k x_k$. If the only acceleration is that due to gravity and the axis $x_3 = z$ points upwards, then $a_1 = a_2 = 0$, $a_3 = -g$, and one has $\tilde{p} = p + \rho g z$. The forcing term, i.e. the last term in the RHS of Eq. (1.42), becomes $g\beta_T(T-T_0)$ in the equation along z $(i = 3)$ and zero elsewhere; note that $g\beta_T(T-T_0)$ is positive where $T > T_0$ and negative elsewhere.

Equation (1.42) is formally identical to the ith Navier–Stokes equation for a constant-property fluid, apart from the different meaning of the pressure term and from the expression of the body force, which is now a function of temperature. Consistently, Eq. (1.42) can be coupled with the constant-density form of the continuity Eq. (1.9) and with the constant-property form of the energy Eq. (1.39) in order to solve a free or mixed convection problem.

Chapter 2
Properties of Turbulence

I am an old man now, and when I die and go to heaven, there are two matters on which I hope for enlightenment. One is quantum electrodynamics and the other is the turbulent motion of fluids. About the former, I am really rather optimistic
Sir Horace Lamb

Abstract With the aid of examples, some fundamental concepts of hydrodynamic turbulence are discussed: *decomposition* into mean and fluctuating values, *energy cascade*, *Kolmogorov* and *Batchelor scales* and *turbulence spectra*. A sketch is also given of the contemporary view that places turbulence in the framework of the theory of *nonlinear dynamic systems* with *chaotic* behaviour and *strange attractors*. In the final section of the chapter, a heuristic approach to turbulence modelling is described which reproduces many of the essential features of more sophisticated models, including the concepts of eddy viscosity/eddy diffusivity and their dependence on the structure of a turbulent flow.

Keywords Decomposition · Energy cascade · Scalar transport · Turbulence spectra · Coherent structure · Strange attractor

2.1 Direct Simulation and Turbulence Modelling

Our starting point are the continuity, momentum and scalar transport equations for a Newtonian fluid written in conservation form, i.e. Equations (1.7), (1.17) and (1.27), repeated here below for the benefit of the reader:

$$\frac{\partial \rho}{\partial t} + \frac{\partial \rho u_j}{\partial x_j} = 0 \tag{1.7}$$

$$\frac{\partial \rho u_i}{\partial t} + \frac{\partial \rho u_j u_i}{\partial x_j} = -\frac{\partial p^*}{\partial x_i} + \frac{\partial}{\partial x_j}\left[\mu\left(\frac{\partial u_i}{\partial x_j} + \frac{\partial u_j}{\partial x_i}\right)\right] + \rho a_i \tag{1.17}$$

$$\frac{\partial \rho \varphi}{\partial t} + \frac{\partial \rho u_j \varphi}{\partial x_j} = \frac{\partial}{\partial x_j}\left(\Gamma \frac{\partial \varphi}{\partial x_j}\right) + S_\varphi \tag{1.27}$$

© The Author(s), under exclusive license to Springer Nature Switzerland AG 2022
M. Ciofalo, *Thermofluid Dynamics of Turbulent Flows*, UNIPA Springer Series,
https://doi.org/10.1007/978-3-030-81078-8_2

Equation (1.27) becomes the energy equation if the scalar φ is the specific enthalpy. Of course, Eq. (1.27) does not apply if the problem is purely hydrodynamic, and applies, but has no effect on the previous equations, if the fluid's properties do not change with the scalar φ (e.g., with temperature).

It has been recognized for a long time (Spalding 1978) that the above equations, provided with suitable boundary and initial conditions, describe to a practically exact extent the behaviour of the fluid both in laminar and in turbulent conditions. Therefore, the extremely complex and apparently stochastic nature of turbulence does not arise from an inadequate mathematical description of the problem nor from external sources of noise (e.g., fluctuations in the boundary conditions), but is an intrinsic property of some solutions of the governing equations themselves.

From the above remarks is follows that, under conditions in which a physical system exhibits a turbulent behaviour, a direct solution (numerical or, in principle, even analytic) of Eqs. (1.7), (1.17) and (1.27), provided it is *sufficiently accurate*, will exhibit the same behaviour (and vice versa). This "brute force" approach, known as Direct Numerical Simulation of turbulence (DNS), requires a great amount of computational resources and is applicable only to a limited class of problems.

The heart of the problem, in fact, lies in the conditions that must be satisfied for a solution to be "sufficiently accurate". As will be discussed in Chap. 3, the crucial constraint is the spatio-temporal resolution necessary for the mechanical energy transfer from the largest to the smallest (so called dissipative) scales of the flow field to be adequately simulated. Many problems of practical interest are not amenable (and will not be for a long time, despite the rapid increment of computing power) to direct numerical simulation, at least with a reasonable amount of resources. This continues and will continue to motivate the recourse to *turbulence models*.

Applying a turbulence model to the study of a fluid dynamics problem amounts to giving up the detailed simulation of the fluid's behaviour, replacing it with an equivalent fluid, generally non-Newtonian (Speziale 1985), defined by suitable constitutive equations, which, under the same boundary conditions, forcing terms etc., exhibits a more regular (and thus more easily predictable) behaviour which, however, can be regarded as a spatially and temporally "smoothed" version of the real system.

There are two main approaches to turbulence modelling. One, called *Large Eddy Simulation* (LES), rests on spatial filtering and requires so called *subgrid models*; the other rests, conceptually, on time averaging over an infinite interval, and leads to the so called *Reynolds Averaged Navier Stokes* (RANS) models. The two approaches will be discussed in Chaps. 4 And 5, respectively.

2.2 Decomposition and Fluctuations

Figure 2.1a is a greyscale map of the instantaneous streamwise (x) velocity u in a xy cross section of a plane channel, represented in the inset. It was obtained by DNS in a finite box imposing periodicity along x and z. The irregular spatial distribution of the selected variable, typical of a turbulent flow, is evident.

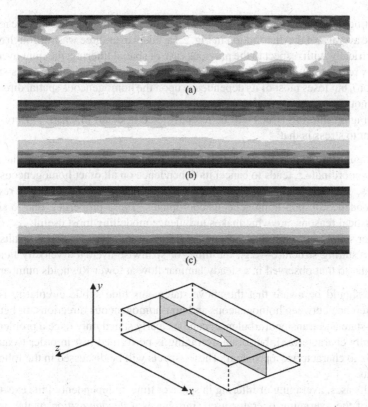

Fig. 2.1 Structure of turbulence in a plane channel, schematically represented in the inset. **a** instantaneous velocity u in a xy plane; **b** instantaneous spatial average of (**a**) along the spanwise direction z; **c** time average of (**a**)

On the complex, multi-scale structure shown in the figure we can apply two distinct *averaging* operations.

The first is *spatial averaging*. Over *homogeneous* directions, defined as those along which any variation is only due to turbulent fluctuations, averaging can be performed over the whole range of the associated spatial variable; in the present example, this is the case of the spanwise and streamwise directions. The instantaneous spanwise average of the temperature distribution in Fig. 2.1a is reported in Fig. 2.1b; the averaged quantity remains a function of the non-homogeneous spatial variable y, but loses almost completely its dependence from the homogeneous streamwise direction x and (not shown) from time t.

As an alternative, *time averaging* can be applied to the structure in Fig. 2.1a, supposedly time-dependent. Time averaging can be performed over an arbitrarily long time (ideally, infinite) only if the problem is statistically stationary; during a transient, or in the presence of non-stochastic oscillations (induced e.g. by a periodic variation of the boundary conditions, as it occurs in meteorological problems), one can only average over a finite time, intermediate between the time scales of turbulent fluctuations and the time scales of the slow, deterministic changes. Figure 2.1c reports the time average of the distribution in Fig. 2.1a, taken over three wash-through times; symmetrically with respect to the previous case of spatial filtering, the time-averaged quantity remains a function of the non-homogeneous spatial variable y (cross-stream direction), but loses most of its dependence upon the homogeneous spatial directions x and (not shown) z.

Averaging/filtering issues will be re-considered in detail in Chap. 7. Here, what we want to stress is that:

- averaging a quantity, e.g. velocity u, over a homogeneous variable, e.g. the span-wise coordinate z, tends to cancel its dependence on all other homogeneous variables, e.g. the streamwise coordinate x or time t, and smooths also its residual dependence on non-homogeneous coordinates. These properties, due to simple statistical reasons, are what makes turbulence modelling most useful;
- under suitable assumptions, space- and time-averaging yield similar results;
- the resulting structure—e.g., the time- or spanwise-averaged velocity field—is similar to that observed in a steady laminar flow at lower Reynolds number.

One should be aware that the above statements hide subtle circularity issues; the difference between homogeneous and non-homogeneous directions, or between steady-state and transient turbulence, can be clearly stated only once a problem has been fully characterized, but often averaging is performed just in order to make it possible to characterize a problem! These aspects will be discussed in the following Sections.

In all cases, averaging or filtering in space or time, independent of the exact definition of the averaging operator used, introduces a *decomposition* of the generic turbulent quantity $\varphi(\mathbf{x}, t)$ into a *mean* (filtered, resolved) component $\langle \varphi \rangle$ and a *fluctuating* (unfiltered, unresolved) component φ', both, in general, still functions of space and time.

2.3 The Kolmogorov Energy Cascade

A noteworthy feature of Fig. 2.1a is the presence of structures characterized by different spatial scales, going from the geometrical size of the channel down to much smaller details. The presence of these multiple scales is the trademark of turbulence: in a laminar flow, the recognizable structures of the flow field would share the size L of the physical structures present in the domain of interest, e.g. the channel's height.

A fundamental scale of turbulence is the *Kolmogorov scale* η_K, characterizing the smallest spatial structures that can exist in the flow field. In the view of turbulence proposed by Kolmogorov as early as in the 1930s, but still basically valid, the mechanical energy supplied to the fluid at a scale L by external processes (pressure gradients, mechanical stirring etc.) undergoes a progressive degradation through a cascade of scales decreasing from L to η_K, when it is eventually dissipated (i.e., converted into heat) by viscous friction (Fig. 2.2).

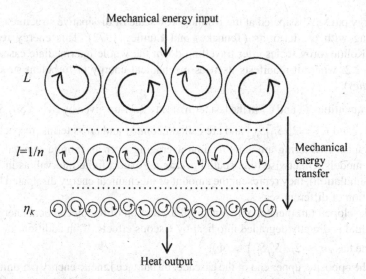

Fig. 2.2 Turbulence energy cascade

Based on purely dimensional arguments one has (Landau and Lifschitz 1959):

$$\eta_K \approx \left(\frac{\nu^3}{\varepsilon}\right)^{1/4} \tag{2.1}$$

in which ν is the fluid's kinematic viscosity (dimensions m^2s^{-1}) and ε is the rate of dissipation of turbulent kinetic energy per unit mass (dimensions m^2s^{-3}).

The quantity ε is related to the dissipation function Φ. For Newtonian incompressible fluids, Φ can be expressed in tensorial notation as:

$$\Phi = \tau_{ij}S_{ij} = 2\mu S_{ij}S_{ij} \tag{2.2}$$

in which (see Appendix) $S_{ij} = \frac{1}{2}(\partial u_i/\partial x_j + \partial u_j/\partial x_i)$ is the strain rate tensor and $\tau_{ij} = 2\mu S_{ij}$ is the traceless part of the viscous stress tensor. Therefore, the filtered value of Φ is $\langle\Phi\rangle = 2\mu\langle S'_{ij}S'_{ij}\rangle$.

By explicitly decomposing S_{ij} as $\langle S_{ij}\rangle + S_{ij}'$, Eq. (2.2) becomes:

$$\langle\Phi\rangle = 2\mu\langle\Phi\rangle\langle S_{ij}\rangle\langle S_{ij}\rangle + 2\mu\langle S'_{ij}S'_{ij}\rangle + 2\mu(L_S + C_S) \tag{2.3}$$

in which the first two terms at the right hand side respectively account for:

- energy dissipated per unit volume and unit time (p.u.v.t.) by direct interaction of the viscous stresses with the filtered (large scale) flow field;

- energy p.u.v.t. dissipated at the Kolmogorov scale of dissipative structures, coinciding with the term $\rho\varepsilon$ (Tennekes and Lumley 1972). This energy reaches
  the Kolmogorov scales after travelling down the whole intermediate cascade in
  Fig. 2.2, where it manifests itself as mechanical energy of turbulent structures
  (*eddies*).

The quantities L_S and C_S in the last term of Eq. (2.3) are given by $L_S = \langle\langle S_{ij}\rangle\langle S_{ij}\rangle\rangle -$
$\langle S_{ij}\rangle\langle S_{ij}\rangle$ and $C_S = 2\langle\langle S_{ij}\rangle S'_{ij}\rangle$ and are called *Leonard* and *cross* terms, respectively.
They vanish if filtering amounts to time-averaging over an infinite interval, as in
RANS models. Otherwise (e.g. spatial averaging over a finite interval, as in Large
Eddy Simulation), they represent the amount of mechanical energy dissipated by the
interaction of different scales.

In developed turbulence, only a negligible fraction of the mechanical energy fed
to the fluid is directly degraded into heat by viscous effects. If, in addition, $L_S = C_S$
$= 0$, one has $\rho\varepsilon = 2\mu\langle S'_{ij} S'_{ij}\rangle \approx \langle\Phi\rangle$.

At the opposite, upper end of the cascade, turbulence kinetic energy per unit mass
$k = \frac{1}{2}\langle u'_i u'_i\rangle$ is generated by shear or by body forces (such as the buoyancy forces
acting in free convection).

The rate of production of ρk by shear, P_k, will be derived in Chap. 5 in the context
of RANS turbulence models. The rate of production of ρk by buoyancy, G_k, will be
derived in Chap. 6 in the context of turbulence in natural and mixed convection. In
local equilibrium, i.e. in a statistically stationary state without advective and diffusive
transport, the production and destruction terms of k balance each other: $\partial(\rho k)/\partial t =$
$(P_k + G_k) - \rho\varepsilon = 0$.

2.4 Turbulence Spectra

The transfer of mechanical energy down the cascade of scales is better appreciated
in the context of the *spectral analysis* of turbulence, some simple elements of which
will be discussed here below. In fact, although turbulent structures are present at
all scales between L and η_K, the kinetic energy associated with them is not evenly
distributed through all scales, but exhibits a characteristic *spectral distribution*.

If at least one homogeneous direction x exists (in the sense defined above), the
generic quantity φ will possess, at any instant t and any values of the remaining
spatial coordinates, a distribution along x, which will typically be complicated and
pseudo-casual. The function $\varphi(x)$ can be represented by the *Fourier integral*:

$$\varphi(x) = 2\pi \int_{-\infty}^{+\infty} a(n)e^{-i2\pi nx} dn \qquad (2.4)$$

in which n is the *wavenumber* (dimensionally the inverse of a length, e.g. m^{-1}). The function $a(n)$ is the Fourier transform of $\varphi(x)$ and is, in general, complex. The *spectral density* $Z_{\varphi\varphi}(n)$ of φ is defined as

$$Z_{\varphi\varphi}(n) = \text{constant} \times |a(n)|^2 \qquad (2.5)$$

in which the constant is a normalization factor, chosen, for example, so as to obtain

$$\int_{-\infty}^{+\infty} Z_{\varphi\varphi}(n)dn = \langle \varphi^2 \rangle \qquad (2.6)$$

(brackets indicate here spatial averaging along the chosen homogeneous direction). It is easily demonstrated that the spectral density $Z_{\varphi\varphi}(n)$ is the Fourier transform of the *spatial auto-correlation function* $R_{\varphi\varphi}(x)$, defined as

$$R_{\varphi\varphi}(x) = \langle \varphi(\xi)\varphi(\xi - x) \rangle = \lim_{X \to \infty} \frac{1}{2X} \int_{-X}^{+X} \varphi(\xi)\varphi(\xi - x)d\xi \qquad (2.7)$$

where ξ is an auxiliary variable.

A typical *velocity* spectrum of developed turbulence is shown in Fig. 2.3. The abscissa is wavenumber n (reciprocal of a wavelength Λ, which represents the size of the associated turbulence structures, or *eddies*), the ordinate is the spectral density $Z_{uu}(n)$ relative to the module of velocity. Thus, $Z_{uu}(n)$ represents the kinetic energy of turbulence per unit wavenumber.

Fig. 2.3 Typical spectrum of velocity in developed turbulence

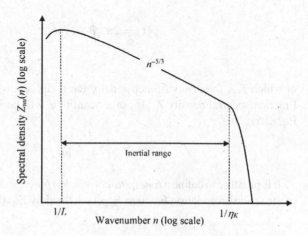

The spectrum exhibits a maximum at the wavenumber associated with the forcing terms ($1/L$), an inertial range in which $Z_{uu}(n)$ is proportional to $n^{-5/3}$ in correspondence with the energy cascade, and a steep roll-off around $1/\eta_K$ in correspondence with the Kolmogorov (dissipative) scale.

This behaviour of $Z_{uu}(n)$ in the inertial range can be derived on the basis of some simplifying assumptions (Lumley 1992). First, suppose the flux (n) of turbulence kinetic energy through the generic wavenumber n (rate of spectral energy transfer from wavenumbers $< n$ to wavenumbers $> n$) to be proportional to the ratio between the kinetic energy associated with n, $u^2(n)$, and the time constant associated with n, $\Lambda(n)/u(n)$. Second, write $u(n) = (n\,Z_{uu}(n))^{1/2}$ and remember that $\Lambda(n) = 1/n$. Finally, observe that $\Sigma(n)$ can be identified with the dissipation ε, which does not depend on the wavenumber; this amounts to assuming that the mechanical energy of turbulence travels down all scales from L to η_K and is dissipated into heat only at the end of the cascade. From the above assumptions, one obtains

$$Z_{uu}(n) \sim \varepsilon^{2/3} n^{-5/3} \tag{2.8}$$

Clearly, the above analysis finds its optimum field of application in numerical methods of the *spectral* type, based on expanding the generic quantity φ in series of eigenfunctions (in particular, in Fourier series). Spectral methods in fluid dynamics are discussed, for example, by Hussaini and Zang (1987).

In principle, the above analysis can be extended to *non homogeneous* spatial directions. However, in this case the spectral distribution of a generic quantity will reflect also its large-scale spatial behaviour and not only the statistics of turbulent spatial fluctuations.

A similar treatment can be applied to the *temporal* distribution of the various flow variables as seen in a fixed point in space (yielding Eulerian spectra). In particular, Eq. (2.4) will be replaced by

$$\varphi(t) = 2\pi \int\limits_{-\infty}^{+\infty} a(f)e^{-i2\pi ft}df \tag{2.9}$$

in which f is *frequency* (dimensionally the reciprocal of a time, e.g. in s^{-1}). The Eulerian spectral density $Z_{\varphi\varphi}(f)$ of a quantity φ will be given by the equivalent of Eq. (2.5):

$$Z_{\varphi\varphi}(f) = |a(f)|^2 \tag{2.10}$$

It is possible to define a *time auto-correlation function* $R_{\varphi\varphi}(t)$, temporal equivalent of the spatial correlation function $R_{\varphi\varphi}(x)$ defined by Eq. (2.7):

$$R_{\varphi\varphi}(t) = \langle \varphi(\tau)\varphi(\tau - t) \rangle = \lim_{T \to \infty} \frac{1}{2T} \int\limits_{-T}^{+T} \varphi(\tau)\varphi(\tau - t)d\tau \qquad (2.11)$$

and the spectral density $Z_{\varphi\varphi}(f)$ is the Fourier transform of $R_{\varphi\varphi}(t)$.

In general, the relation between spatial spectral densities or spatial correlation functions on one side, and their Eulerian temporal counterparts on the other side, is not simple. In some cases Taylor's *frozen turbulence* assumption can be adopted, according to which the temporal distribution observed from a point P fixed in space is equal to the distribution that would be observed if the flow field were *frozen* and translated in front of P with an advective velocity U. One then obtains, for example, a frequency of dissipative eddies (maximum significant frequency in the Eulerian time spectrum) equal to

$$f_K = U/\eta_K \approx U(v^3/\varepsilon)^{-1/4} \qquad (2.12)$$

2.5 Scalar Transport in Turbulence

The transport of a scalar φ is governed by Eq. (1.27). All the general properties and definitions of the previous section, and, in particular, Eqs. (2.4)–(2.7) and (2.9)–(2.11) hold for the spectra of φ as well as for those of velocity. However, the shape of the spectral density curve $Z_{\varphi\varphi}(n)$, which now expresses the variance of φ per unit wavenumber, may be deeply different for a scalar than for velocity, depending on the value of the Schmidt number Sc $= \mu/\Gamma = \nu/D$, ratio of the momentum and scalar diffusivities (Sreenivasan 2019). Note that, when the scalar is specific enthalpy, Sc is commonly denoted as Prandtl number σ.

Figure 2.4 shows three possible behaviours of the scalar spectral density $Z_{\varphi\varphi}$, corresponding to three limiting cases: (a) Sc$\approx$1; (b) Sc$\gg$1; and (c) Sc $\ll$ 1.

(a) For Sc $\approx$ 1 (solid line in Fig. 2.4), the scalar spectral density behaves much as the velocity spectral density. It is characterized by an inertial range in which $Z_{\varphi\varphi} \sim n^{-5/3}$, extending from the wavenumber $1/L$ corresponding to the large scales of the flow to the wavenumber $1/\eta_K$ corresponding to the Kolmogorov scale of dissipative eddies, followed by a steep roll-off. In fact, in this case η_K is also the spatial scale of the smallest "features" (inhomogeneities) in the distribution of φ that can survive diffusion.

Notably, this case includes also heat transfer to fluids having a Prandtl number close to unity (all gases and, approximately, also water and other refrigerants).

(b) For Sc $\gg$ 1 (dash-dot line in Fig. 2.4), the smallest spatial scale of the scalar inhomogeneities is *not* η_K, but the so called *Batchelor scale* η_B equal to $\eta_K/\text{Sc}^{1/2}$, and thus much smaller than the Kolmogorov scale. Therefore, while fluid "lumps" of size η_K kinematically behave as rigid bodies (without significant velocity

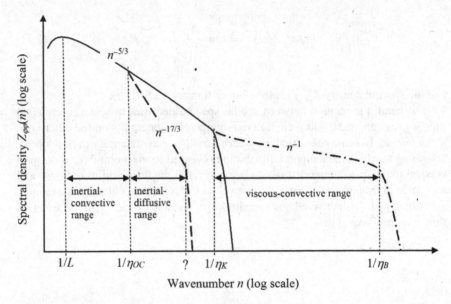

Fig. 2.4 Possible scalar spectra in developed turbulence. Solid line: Sc $\approx$ 1; dash-dot line: Sc $\gg$ 1; broken line: Sc $\ll$ 1

differences between their parts), for Sc $\gg$ 1 they can still accommodate within their volume significant scalar gradients, i.e. smaller "lumps" of fluid of size $\sim \eta_B$ having different values of φ. The spectral density of φ follows the $-5/3$ law up to the wavenumber $1/\eta_K$, and then, instead of falling steeply, changes slope entering what is called the *viscous-convective range*, characterized by a n^{-1} behaviour, to eventually roll-off at $n \approx 1/\eta_B$.

The condition Sc $\gg$ 1 occurs in two classes of phenomena: heat transport in high Prandtl number fluids (e.g. oils and molten polymers), and mass transport in the presence of low-diffusivity solutes (almost all solutes in water exhibit $D \ll \nu$, i.e. Sc $\gg$ 1). In this latter case, the scalar φ is the concentration of the solute. Transport phenomena occurring in the viscous-convective range, i.e. at scales intermediate between the Batchelor and the Kolmogorov lengths, are often called *micromixing* in the mixing community, as opposed to *macromixing* and *mesomixing* phenomena occurring in the inertial range of scales larger than η_K (Baldyga and Pohorecki 1995).

(c) For Sc $\ll$ 1 (broken line in Fig. 2.4), the smallest possible scale at which scalar distribution inhomogeneities can exist without being destroyed by diffusion is larger than the Kolmogorov scale η_K; it is called the *scalar dissipation length*, or *Obukhov-Corrsin length*, is denoted by η_{OC}, and is related to η_K and Sc by $\eta_{OC} = \eta_K/Sc^{3/4}$. In this case, opposite to the previous one Sc $\gg$ 1, one may expect to find "lumps" of fluids that are homogenous as far as the distribution of the scalar φ is concerned, but still exhibit within their volume significant velocity gradients. The spectral density follows the $n^{-5/3}$ behaviour only in the *inertial-convective range* up to the wavenumber $1/\eta_{OC}$, and then changes slope to a much steeper $n^{-17/3}$ behaviour. The roll-off occurs at a wavenumber $> 1/\eta_{OC}$; the spectral behaviour in this range is still unclear.

The condition Sc « 1 occurs in heat transfer with liquid metals (σ « 1) but is rarely encountered in mass transfer problems.

2.6 Coherent Structures

Despite the pseudo-casual nature of turbulent fluctuations, the constraints that the continuity and Navier–Stokes equations impose to the flow field cause it to be highly structured at all scales. Of course, turbulent structures (*eddies*) are not permanent but exhibit a certain mean lifetime. The research on coherent structures has been among the most productive in the field of turbulence studies. Examples include:

- the vortex structures in the near-wall region of a turbulent channel flow. Figure 2.5 schematically shows a *low-speed streak* and, above it, consecutive stages in the evolution of an initial *roll* vortex, through the *horseshoe* and *hairpin* configurations until its breakdown, associated with a *turbulence burst*. The burst is preceded by an *ejection* event that transports fluid away from the wall and followed by a *sweep-inrush* event that transports fluid towards the wall. The sequence repeats itself cyclically, of course in an approximate and not in a strictly periodic fashion, and is thought to be responsible for most of the fluid-wall shear stress (i.e., friction and pressure loss). The low-speed streaks were first observed by Kline et al. (1967); subsequent experimental investigations and direct numerical simulations have progressively elucidated the nature of the other near-wall structures and their role in the wall turbulence cycle (Barrett 1990; Jiménez and Pinelli 1999).

- the vorticity tubes that populate the wake of an obstacle, as observed, for example, by Schröder et al. (2020), Fig. 2.6;

Fig. 2.5 Wall turbulence structures

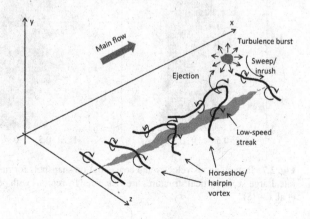

Fig. 2.6 Wake of a surface-mounted cube for $U_\infty = 0.4$ m/s (laminar in-flow): instantaneous vorticity iso-surfaces ($|\omega|=170$ s^{-1}) obtained by a particle tracking experimental technique. Reprinted from Schröder et al. (2020)

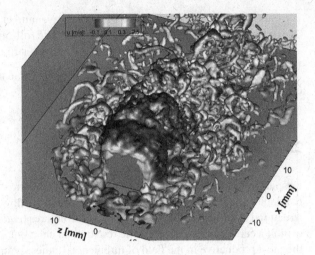

- the toroidal vortices detected by Dulin et al. (2018) at the edge of turbulent jets, Fig. 2.7.

Further examples include the quasi-two dimensional, highly organized, vortices first observed by Brown and Roshko (1974) at the turbulent interface between planar jets; the coherent structures present even in homogeneous isotropic turbulence (Yuan et al. 2006); the large structures of planetary atmospheres, of which Jupiter's Great Red Spot, first observed by Cassini in 1665, is probably the most persistent.

The research on coherent structures has been among the most productive in the field of turbulence studies. Among other merits, it has solicited the development of experimental and numerical techniques to extract patterns from a vast and noisy background of time-dependent data. These include, for example, conditional sampling

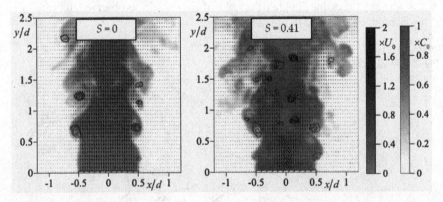

Fig. 2.7 Instantaneous velocity and concentration snapshots for turbulent jets with different swirl rate. Large-scale vortical structures are visualized by regions with positive Q-criterion. From Dulin et al. (2018)

(Antonia 1981), Proper Orthogonal Decomposition (Berkooz et al. 1993), and the use of the Q-parameter (see Section A.3 in the Appendix) to identify vortical structures. A recently proposed technique for the processing of experimental particle images, dubbed "Shake-The-Box" (Schanz et al. 2016), has allowed the tracking of individual particles at numbers of up to hundreds of thousands per time step.

2.7 Three-Dimensional Versus Two-Dimensional Turbulence

Turbulence is intrinsically *three-dimensional*. This fact is discussed in a particularly clear way by Bradshaw (1978). He focusses on the phenomenon of *vortex stretching* by large-scale velocity gradients, Fig. 2.8.

The large-scale velocity gradient existing in a region of the flow field stretches and thins an initially existing vorticity tube. As a first approximation, when the length l_ω of the vorticity tube increases, its radius r and its angular velocity Ω must vary so as to ensure conservation of both mass and angular momentum, proportional to $R^2 l_\omega$ and $R^4 l_\omega \Omega$, respectively. As a consequence, the kinetic energy of the vortex, which is proportional to $R^4 l_\omega \Omega^2$, must increase, and does this at the expense of the energy of the large-scale motion. Therefore, vortex stretching transfers energy from large to smaller scales, thus playing a major role in the energy cascade.

This phenomenon is intrinsically 3-D and cannot exist in a 2-D problem, in which vorticity is orthogonal to the plane of the flow so that vortices cannot be stretched.

Thus, one may expect that, in two dimensions, the energy cascade develops in a different way than in three, Kolmogorov spectra do not apply and, in general, the features of turbulence will differ from those of real-world, 3D, turbulence. Theoretical studies, based on the concept of *enstrophy*, or vorticity variance (Kraichnan 1967; Lesieur 1990), predict for the spectral density of two-dimensional isotropic turbulence a $n^{-5/3}$ trend for $n < n_{in}$ and a steeper n^{-3} trend for $n > n_{in}$, n_{in} being the wavenumber at which mechanical energy is injected into the fluid. Moreover,

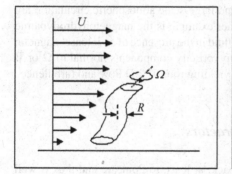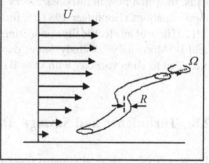

Fig. 2.8 Vortex stretching by large-scale velocity gradients in a three-dimensional flow

Fig. 2.9 Streamlines at consecutive instants for two-dimensional turbulent free convection in an internally heated, low Prandtl number fluid enclosed in a rectangular cavity. Solid lines: clockwise circulation; broken lines: counter-clockwise circulation. From results of Di Piazza and Ciofalo (2000)

2D turbulence exhibits the phenomenon called *inverse energy cascade* (mechanical energy transfer from smaller to larger scales). Finally, the predicted dissipative scale is of the order of $(v^3/\beta_E)^{1/6}$, in which β_E is the rate of dissipation of enstrophy, or vorticity variance (analogue of the quantity ε which represents the rate of change of turbulence kinetic energy k, or velocity variance), and thus differs in general from Kolmogorov's scale $(v^3/\varepsilon)^{1/4}$.

In view of the differences described above, two-dimensional direct numerical simulations of physically three-dimensional turbulent problems should be regarded with caution. Yet, because of their moderate requirement of computational resources, they were fairly common until some years ago (Nobile 1996). For example, Fig. 2.9 shows consecutive streamline snapshots for turbulent free convection of an internally heated liquid metal in a rectangular cavity (Di Piazza and Ciofalo 2000).

Physical situations do exist, in which the structure of turbulence, and, more generally, of the flow field is (quasi) 2-D. An important instance is the flow between plane parallel walls in the presence of rotation with angular velocity Ω orthogonal to the walls, of which flow in turbomachinery and large-scale atmospheric circulation are approximations (Hopfinger 1987); a further example is the magnetohydrodynamic (MHD) flow of an electrically conducting fluid in the presence of an intense magnetic field **B** (Moreau 1990). Body forces damp vorticity components normal to Ω or **B** (i.e., tend to align vorticity with Ω or **B**), yielding (quasi) 2-D flow and turbulence.

2.8   Turbulence and *Strange Attractors*

The theory of dynamical systems dates back, at least, to Poincarè and thus is well more than a century old. However, only in the last three or four decades the advent of sufficiently powerful computers has made the direct investigation of the structure and behaviour of large nonlinear dynamical systems feasible.

A *discrete variable, continuous time* dynamical systems consists of N quantities $x_1 \ldots x_N$ which vary in time according to evolution equations (which in general are non-linear) of the form

$$\partial x_i / \partial t = F(x_1 \ldots x_N, t) \qquad (2.13)$$

($i = 1 \ldots N$). A fluid in turbulent motion is a *continuous variable* dynamical system and thus is characterized by *infinite* degrees of freedom. However, any discrete, approximate numerical representation of the problem (based, for example, on N_p grid points) can be described as the time evolution of a finite number of variables (for example, the $N = 5 N_p$ variables representing the three velocity components, pressure and density at each grid point); as it will be shown later, a numerical simulation based on N_p grid points can be regarded as complete provided the size of the cells in the computational grid is less than the Kolmogorov scale η_K characteristic of the flow.

In the *phase space* defined by the N axes $x_1 \ldots x_N$ the state of the dnamical system is represented by a single point, and its evolution in time describes a line (*trajectory*). If the system is *dissipative*, i.e. includes a mechanism for degradation of mechanical energy into heat (as it necessarily happens in fluid dynamics due to the presence of viscous terms), for any initial condition I the system's trajectory will eventually end in a subset of phase space, having dimensionality lower than the whole space and thus necessarily of zero measure in it, which is called the system's *attractor*. Many such attractors can co-exist, each characterized by an *attraction basin* (set of all possible initial configurations for which the system eventually falls into that attractor). For more rigorous definitions, see e.g. Lanford (1981).

Long before 1960, classic analytical mechanics had identified a number of simple attractors, schematically represented in graphs a–c of Fig. 2.10: *fixed points* (P), characteristic of *steady states*; *closed orbits*, or limit cycles (C), characteristic of *periodic states*; and *open orbits* confined on a 2-torus (T), characteristic of *quasi-periodic states* exhibiting two incommensurate frequencies.

Lorenz (1963) and other scientists discovered simple non-linear dynamical systems whose attractors do *not* fall into any of the above types; they are *fractal* mathematical structures, infinitely complex although of zero measure in phase space, which were named strange *attractors*. An example (S) is represented (*very* schematically!) in graph (d) of Fig. 2.10.

The evolution of a dynamical system on a strange attractors exhibits peculiar properties, among which *sensitivity to initial conditions*: trajectories starting from two arbitrarily close points rapidly diverge, giving rise to macroscopically different evolution histories. A closely connected concept is *deterministic chaos*.

All these properties can be more rigorously characterized only by introducing such concepts as the attractor's *fractal dimension* and the *Ljapunov* exponents of the trajectories. Despite the hard underlying maths, strange attractors have been the

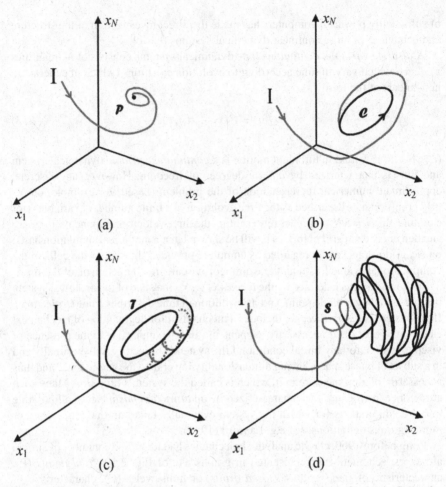

Fig. 2.10 Attractors in the phase space of a dissipative dynamical system. **a** fixed point P (steady state); **b** closed orbit, or limit cycle, C (periodic regime); **c** open orbit T confined on a 2-torus (quasi-periodic regime); **d** *strange attractor* S (chaotic, or turbulent, regime)

subject, especially in the years 1980s and 1990s, of an explosive literature, not only specialistic but also popular, largely because of their aesthetic appeal.

Starting from a seminal paper by Ruelle and Takens (1971), hydrodynamic turbulence has been re-reinterpreted in the light of this new view of dynamical systems. The basic intuitions are:

- the object responsible for turbulence is an attractor of relatively small dimensionality, embedded in the high-dimensional phase space representing all possible flow fields;
- the system's evolution on the attractor depends sensitively on initial conditions;

- this dependence accounts for the chaotic and apparently stochastic nature of turbulent flows.

Ruelle and Takens hoped that, in the foothpath of Lorenz's studies, a generic turbulent problem could be interpreted on the basis of a reasonably small number of degrees of freedom, so as to allow many of the flow features to be deduced from the direct investigation of the attractor's structure.

Unfortunately, these expectations have largely gone disappointed. Témam (1988) predicted that the strange attractors of turbulence would be of huge dimensionality. Keefe et al. (1992), starting from direct numerical simulations of turbulent plane Poiseuille flow at Reynolds numbers of ∼ 4000 and processing the results in the light of the Ljapunov exponent theory, concluded that the attractor of the corresponding dynamical system must possess a dimensionality of 352 (!).

These results confirmed that turbulence in channel flows really is *deterministic chaos*, related to the existence of a strange attractor; but, at the same time, barred (due to the very large dimensionality of this latter) any hope of modelling the problem on the basis of *few* degrees of freedom, or deducing significant flow properties from a direct analysis of the attractor.

2.9 A Heuristic Approach to Turbulence Modelling

Despite the great complexity of turbulence, of which the present chapter should have provided ample evidence, in this final Section it will be shown that even elementary heuristic arguments, based on an extremely simplified representation of a turbulent flow, are sufficient to derive a "miniature" turbulence model which, however, exhibits many important features of better established ones, including the concepts of turbulent viscosity and diffusivity.

Consider a fluid of density ρ flowing along x over a flat solid wall, and let the "base" (averaged or filtered) velocity $\langle u \rangle$ be a function only, or mainly, of the coordinate y normal to this wall, so that $\langle u \rangle = \langle u \rangle (y)$ (Fig. 2.11). Suppose the flow to be turbulent, and assume that turbulence consists of an array of identical vortices, or *eddies*, of radius $R = l/2$ and angular velocity ω, distributed along the wall with a mean spacing P between one another and dragged by the base velocity field.

Under these assumptions, each eddy rotates with a peripheral velocity $v = \omega R$ and completes a revolution in a period $t_{per} = 2\pi/\omega$. For dimensional consistence, it is convenient to assume that the whole fluid domain has a finite extent W in the spanwise direction z orthogonal to the plane xy of the figure.

Let $m = \pi R^2 W \rho$ be the mass of an eddy. At each half revolution (180° turn), a fluid parcel of mass $m/2$ and streamwise velocity $\langle u \rangle_2 = \langle u \rangle (y_2)$ (labeled "faster fluid" in Fig. 2.11) is transported into a fluid stream closer the wall where the base velocity is $\langle u \rangle_1 = \langle u \rangle (y_1)$. At the same time, a fluid parcel of identical mass $m/2$ and streamwise velocity $\langle u \rangle_1 = \langle u \rangle (y_1)$ (labeled "slower fluid" in Fig. 2.11) is transported into a fluid stream farther from the wall where the base velocity is $\langle u \rangle_2 = \langle u \rangle (y_2)$.

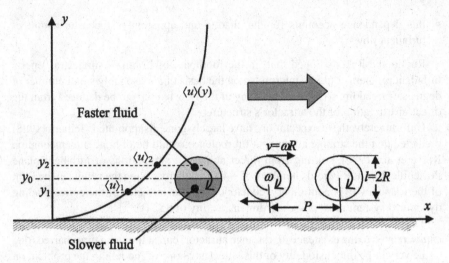

Fig. 2.11 Heuristic model for the derivation of the eddy viscosity concept and of its dependence on the turbulence scales

The net effect is a transfer of a longitudinal momentum $(m/2)(\langle u \rangle_2 - \langle u \rangle_1)$ across the plane $y = y_0$ containing the eddy axes.

Dividing this momentum (impulse) by the time $t_{per}/2 = \pi/\omega$ of half revolution, one obtains the force f_1 (directed along the positive direction of the x axis) that, for each eddy, the fluid above y_0 exerts on the fluid below. Using the above definitions and observing that $\langle u \rangle_2 - \langle u \rangle_1 \approx (l/2)(\mathrm{d}\langle u \rangle/\mathrm{d}y)$, one has

$$f_1 = \frac{\left(\frac{1}{2}\pi R^2 W \rho \right) \times (l/2)}{\pi/\omega} \frac{\mathrm{d}\langle u \rangle}{\mathrm{d}y} = \frac{1}{16}\rho W \omega l^3 \frac{\mathrm{d}\langle u \rangle}{\mathrm{d}y} \qquad (2.14)$$

Dividing f_1 by the area of the plane $y = y_0$ associated with each eddy, WP, one obtains the tangential force per unit area, or *shear stress*, exerted by the fluid above y_0 on the fluid below as a consequence of turbulence:

$$\tau_t = \frac{\rho W \omega l^3}{16WP} \frac{\mathrm{d}\langle u \rangle}{\mathrm{d}y} = \left(\frac{1}{8}\frac{l}{P} \right) \rho l v \frac{\mathrm{d}\langle u \rangle}{\mathrm{d}y} \qquad (2.15)$$

Remembering that, in steady parallel flow, one can write Newton's law of viscosity as $\tau = \mu(\mathrm{d}u/\mathrm{d}y)$, Eq. (2.15) can be written in the similar form $\tau_t = \mu_t(\mathrm{d}\langle u \rangle/\mathrm{d}y)$ provided the turbulent viscosity μ_t is defined as

$$\mu_t = \left(\frac{1}{8}\frac{l}{P} \right) \rho l v \qquad (2.16)$$

i.e., as the product of a purely numerical "turbulence density" coefficient $l/(8P)$, which can range from 0 to 0.125, by density ρ, eddy size l and eddy peripheral velocity v (identifiable with the root mean square value of the velocity fluctuations).

Despite its simplicity, the heuristic model described above leads to two important conclusions:

- the effects of turbulence amount to the appearance of a turbulent stress that adds itself to the viscous stress and, like this latter, is proportional to the base velocity gradient, so that a turbulent, or eddy, viscosity μ_t can be defined;
- the eddy viscosity, apart from the density factor and from a numerical constant (presumably of order 0.1), is proportional to two distinct features of turbulence: the size l of the turbulent eddies and their peripheral velocity v.

Besides the turbulence scales defined above, i.e. l, v, ω, further quantities can be built from them, e.g. $k \approx v^2$ (turbulent kinetic energy per unit mass), $\varepsilon \approx k\omega$ (turbulent kinetic energy dissipation rate per unit mass) and others. Of all these quantities, once two independent ones are chosen, all others remain determined by the existing relations between the various scales; therefore, even at this crude level of description, turbulence is characterized by *two degrees of freedom*, e.g. k and l, k and ε or k and ω, meaning that μ_t can be expressed as a function of two of the scale variables. This is the main reason why most eddy-viscosity RANS model are based on two transport equations. The choice of which couple of quantities to use for the computation of μ_t is the basis for the buildup of alternative two-equation RANS models, as will be discussed in Chap. 6.

Similar considerations can be applied to the transport of a scalar quantity, e.g. specific enthalpy $c_p T$, to derive the concept of *eddy diffusivity* and its dependence on the turbulence features.

With reference to Fig. 2.12, at each half revolution a fluid parcel of mass $m/2$ and temperature $\langle T \rangle_2 = \langle T \rangle(y_2)$ (labeled "hotter fluid" in Fig. 2.12) is transported into a fluid stream closer the wall where the base temperature is $\langle T \rangle_1 = \langle T \rangle (y_1)$. At the same time, a fluid parcel of identical mass $m/2$ and temperature $\langle T \rangle_1 = \langle T \rangle(y_1)$ (labeled "colder fluid" in Fig. 2.11) is transported into a fluid stream farther from the wall where the base temperature is $\langle T \rangle_2 = \langle T \rangle(y_2)$. The net effect is a transfer of enthalpy $(m/2)c_p(\langle T \rangle_2 - \langle T \rangle_1)$ across the plane $y = y_0$ containing the eddy axes.

Dividing this enthalpy (thermal energy) by the time $t_{per}/2 = \pi/\omega$ for half revolution, one obtains the thermal power Q_1 that, for each eddy, the fluid above y_0 transfers the fluid below. Using the above definitions and observing that $\langle T \rangle_2 - \langle T \rangle_1 \approx (l/2)(d\langle T \rangle/dy)$, one has

$$Q_1 = \frac{\left(\frac{1}{2}\pi R^2 W \rho\right) \times c_p(l/2)}{\pi/\omega} \frac{d\langle T \rangle}{dy} = \frac{1}{16} \rho W \omega l^3 c_p \frac{d\langle T \rangle}{dy} \tag{2.17}$$

Dividing Q_1 by the area of the plane $y = y_0$ associated with each eddy, WP, one obtains the heat flux per unit area transferred across the plane $y = y_0$ as a consequence of turbulence:

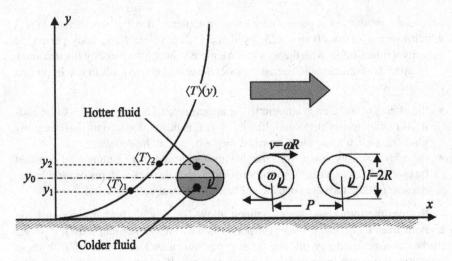

Fig. 2.12 Heuristic model for the derivation of the eddy diffusivity concept and of its dependence on the turbulence scales

$$q_t'' = \frac{\rho W \omega l^3 c_p}{16 W P}\frac{d\langle T\rangle}{dy} = \left(\frac{1}{8}\frac{l}{P}\right)\rho l v c_p \frac{d\langle T\rangle}{dy} \tag{2.18}$$

Remembering Fourier's law of conductive heat transfer $q = \Gamma c_p(dT/dy)$, Eq. (2.18) can be written in the similar form $q_t = \Gamma_t c_p(d\langle T\rangle/dy)$ provided the turbulent diffusivity Γ_t is defined as

$$\Gamma_t = \left(\frac{1}{8}\frac{l}{P}\right)\rho l v \tag{2.19}$$

The comparison of Eq. (2.19) with Eq. (2.16) shows that the eddy diffusivity Γ_t and the eddy viscosity μ_t are identical:

$$\Gamma_t = \mu_t \tag{2.20}$$

By analogy with the definition of the (molecular) Prandtl number of a fluid as the ratio of the viscosity to the thermal diffusivity ($\sigma = \mu/\Gamma$), Eq. (2.20) can be expressed as the statement that the ratio of the eddy viscosity μ_t to the eddy diffusivity Γ_t, i.e. the *turbulent Prandtl number*, is equal to 1 ($\sigma_t = \mu_t/\Gamma_t = 1$).

An analysis a little more sophisticated than the crude heuristic approach followed here leads to conclude that the turbulent Prandtl number in channel and boundary layer flows should actually range between 0.8 and 0.9. What is important to observe is that unlike the (molecular) Prandtl number σ, which varies by orders of magnitude from fluid to fluid (from ~ 10^{-3} in liquid metals to 10^4 and more in dense oil, glycerol or molten polymers), the turbulent Prandtl number σ_t does *not* depend on the fluid's thermophysical properties.

Once again, an extremely simplified model leads to two correct conclusions of outstanding importance:

- the effects of turbulence on heat transport amount to the appearance of a turbulent heat flux that adds itself to the conductive flux and, like this latter, is proportional to the base temperature gradient, so that a turbulent, or eddy, diffusivity Γ_t can be defined;
- the eddy diffusivity is proportional to the eddy viscosity via a turbulent Prandtl number of unity order (and thus, like μ_t, is proportional to two distinct features of turbulence, the size l and the peripheral velocity v of the eddies).

The extension of the above results to different scalars φ, e.g. concentrations, is straightforward. In this case, however, the term *turbulent Schmidt number* will be used instead of *turbulent Prandtl number*, by analogy with the corresponding molecular quantity.

A further simplification is obtained by adopting Prandtl's *mixing length* view (Prandtl 1925). According to this view, the length and velocity scales of turbulence l, v are not independent, but rather related by

$$v = l|d\langle u\rangle/dy|$$ (2.21)

Here, the length scale l is regarded more as a *penetration length* of the velocity fluctuations than as a physical eddy size. Independent of the interpretation to be given to Eq. (2.21), its adoption clearly reduces from two to a single one (l) the degrees of freedom of turbulence. The expression of the eddy viscosity, Eq. (2.16), becomes

$$\mu_t = \text{constant} \times \rho l^2 \left|\frac{dU}{dy}\right|$$ (2.22)

Of course, the quantity l (*Prandtl mixing length*) remains to be modelled. Prandtl himself assumed l to be proportional to the distance y from the wall, while von Karman (1931) suggested to compute l as $(\partial\langle u\rangle/\partial y)/\partial^2\langle u\rangle/\partial y^2$. Van Driest (1956) proposed to account for the reduction of the eddy size caused by the proximity of the wall multiplying the length scale l in Eq. (2.22) by a damping factor

$$f_\mu = 1 - \exp(-y^+/A^+)$$ (2.23)

in which y^+ is the distance from the nearest wall in "wall units" v/u_τ and $A^+ = 25$.

The concept of mixing length is so primitive that it applies both to RANS models and to Large Eddy Simulation (where one speaks of a *subgrid* viscosity rather than of an *eddy* viscosity). It will be recalled when required in discussing these families of models, Chaps. 4 and 5.

References

Antonia RA (1981) Conditional sampling in turbulence measurement. Annu Rev Fluid Mech 13:131–156

Baldyga J, Pohorecki R (1995) Turbulent micromixing in chemical reactors—a review. Chem Eng J 58:183–195

Barrett J (1990) Shark skin inspires new surface technology. Eureka on Campus 2(2):14–16

Berkooz G, Holmes P, Lumley JL (1993) The proper orthogonal decomposition in the analysis of turbulent flows. Annu Rev Fluid Mech 25:539–575

Bradshaw P (1978) Introduction. In: Bradshaw P (ed) Turbulence. Springer, Berlin, pp 1–44

Brown GL, Roshko A (1974) On density effects and large structure in turbulent mixing layers. J Fluid Mech 64:775–816

Di Piazza I, Ciofalo M (2000) Low-Prandtl number natural convection in volumetrically heated rectangular enclosures—I. slender cavity, AR=4. Int J Heat Mass Transf 43:3027–3051

Dulin V, Lobasov A, Markovich D, Alekseenko S (2018) Coherent structures in the near field of swirling turbulent jets and flames investigated by PIV and PLIF. In: Boushaki T (ed) Swirling flows and flames, chapter 2. Intechopen

Hopfinger EJ (1987) Turbulence in stratified fluids: a review. J Geophys Res 92:5287–5303

Hussaini MY, Zang TA (1987) Spectral methods in fluid dynamics. Annu Rev Fluid Mech 19:339–367

Jiménez J, Pinelli A (1999) The autonomous cycle of near-wall turbulence. J Fluid Mech 389:335–359

Keefe L, Moin P, Kim J (1992) The dimensions of attractors underlying periodic turbulent Poiseuille flow. J Fluid Mech 242:1–29

Kline SJ, Reynolds WC, Schraub FA, Rundstadler PW (1967) The structure of turbulent boundary layers. J Fluid Mech 30:741–773

Kraichnan RH (1967) Inertial ranges in two-dimensional turbulence. Phys Fluids 10:1417–1423

Landau LD, Lifshitz EM (1959) Fluid mechanics. Pergamon Press, Reading, MA

Lanford OE (1981) Strange attractors and turbulence. In: Swinney HL, Gollub JP (eds) Hydrodynamic Instabilities and the transition to turbulence. Springer, Berlin, pp 7–26

Lesieur M (1990) Turbulence in Fluids. Kluwer Academic Publishers, Dordrecht

Lorenz EN (1963) Deterministic nonperiodic flow. J Atmos Sci 20:130–141

Lumley JL (1992) Some comments on turbulence. Phys Fluids A 4:203–211

Moreau R (1990) Magneto-hydrodynamic turbulence. Kluwer, Dordrecht

Nobile E (1996) Simulation of time-dependent flow in cavities with the additive-correction multigrid method. Part I: mathematical formulation; Part II: applications. Numer Heat Transf 30:341–350 and 351–370

Prandtl L (1925) Bericht über Untersuchungen zur ausgebildeten Turbulenz (Report on investigation about developed turbulence). Z Angew Math Mech 5(1):136–139

Ruelle D, Takens F (1971) On the nature of turbulence. Commun Math Phys 20:167–192

Schanz D, Gesemann S, Schröder A (2016) Shake-The-Box: lagrangian particle tracking at high particle densities. Exp Fluids 57:70

Schröder A, Willert C, Schanz D, Geisler R, Jahn T, Gallas Q, Leclaire B (2020) The flow around a surface mounted cube: a characterization by time-resolved PIV, 3D Shake-The-Box and LBM simulation. Exp Fluids 61:189

Spalding DB (1978) Discussion on "Turbulence models for heat transfer". In: Proceedings 8th international heat transfer conference, Toronto, Canada, vol 8, p 8

Speziale CG (1985) Galilean invariance of subgrid-scale stress models in the large-eddy simulation of turbulence. J Fluid Mech 156:55–62

Sreenivasan KR (2019) Turbulent mixing: a perspective. Proc Nat Acad Sci U S A (PNAS) 116(37):18175–18183

Témam R (1988) Infinite-dimensional dynamical systems in mechanics and physics. Springer, New York

Tennekes H, Lumley JL (1972) A first course in turbulence. MIT Press, Cambridge, MA

Van Driest ER (1956) On turbulent flow near a wall. J Aero Sci 23:1007–1011

von Karman T (1931) Mechanical similitude and turbulence. NACA Technical Memorandum NACA-TM-611

Yuan X, Nguyen MX, Chen B, Porter DH (2006) HDR VolVis: High dynamic range volume visualization. IEEE Trans Visual Comp Graph 12(4):433–445

Chapter 3
Direct Numerical Simulation (DNS)

Let him but copy what in you is writ, Not making worse what
Nature made so clear
William Shakespeare, Sonnet LXXXIV

Abstract The spatial and temporal resolution requirements for the direct numerical simulation of a turbulent flow problem are derived. The computational cost of a simulation is discussed in the light of contemporary advances in computing performances, and an example of very high resolution numerical simulation is provided.

Keywords Direct numerical simulation · Reynolds number · Computational grid · Kolmogorov scale · Parallel computing · Plane channel

3.1 Direct Numerical Simulation and Spatio-Temporal Resolution

In any fluid dynamics problem, the accurate numerical solution of the governing equations requires the computational domain to be covered by a three-dimensional grid whose pitch Δx (spatial discretization) must be small enough to resolve the smallest significant spatial features of the flow field. Also, in time-dependent problems, the simulation must be performed using time steps Δt (time discretization) small enough to resolve the time-dependent behaviour of the various quantities.

If the Reynolds number (ratio of inertial to viscous forces) is sufficiently small, the flow is laminar and, as anticipated in Sect. 2.3, the size of all significant features of the flow field is of the same order as that L of the physical structures contained in the computational domain (pipe diameter, obstacle size etc.). Moreover, if the boundary conditions and the forcing terms do not change with time, the problem always admits one or more steady or periodic solutions (possibly following an initial transient). Finally, the existence of spatial symmetries generally translates into corresponding symmetries of the flow field and allows the simulation to be performed in only a part of the physical domain of interest, e.g. in two dimensions rather than three, thus greatly speeding up the necessary computations. Therefore, in this case it is generally possible to achieve a sufficient spatial and temporal resolution and to obtain solutions which are independent of the specific discretization adopted (grid independence and time step independence).

For higher Reynolds numbers, the flow becomes turbulent. In this case, the flow is always time-dependent (even in the presence of constant boundary conditions and forcing terms), depends critically on the initial conditions, remains fully three-dimensional and does not instantaneously exhibit any spatial symmetry (even in the presence of symmetric domain shapes and boundary conditions or forcing term distributions). What is more important, as discussed throughout Chap. 2, the spatial features of the flow field extend over an interval of scales ranging from the size of the whole computational domain down to that of the dissipative eddies (Kolmogorov scale). This range increases with the Reynolds number and, for developed turbulence, may cover several orders of magnitude. As stated in Sect. 2.1, the governing equations remain an accurate description of the problem's physics, but their numerical solution presents serious difficulties. These will be quantitatively assessed in the following Sect. 3.1 for a specific problem, i.e. fully developed turbulent flow between plane parallel plates (turbulent plane Poiseuille flow).

3.2　Resolution Requirements for Turbulent Plane Poiseuille Flow

Consider the turbulent flow in a channel limited by plane parallel walls and indefinite in the streamwise and spanwise directions. In order to study this system, it is possible to choose as the computational domain a portion of channel in which it can be assumed that the velocity field repeats itself periodically. In order to make the solution statistically independent from the dimensions of the domain, these must be such that the correlations between variables located at the opposite sides are negligible; this amounts to choosing dimensions sufficiently larger than the largest expected turbulent structures (eddies).

More precisely, let the x axis be aligned with the direction of the mean flow, the y axis orthogonal to the walls and the z axis oriented spanwise, and let δ be the channel's half height (Fig. 3.1). The dimensions of the computational domain ("box") can be expressed as $L_x = K_x \delta$, $L_y = 2\delta$, $L_z = K_z \delta$. Classic experimental results (Comte-Bellot 1963) show that the statistical correlations between instantaneous quantities become negligible at distances of $\sim 4\delta$ in the longitudinal direction (x) and $\sim 2\delta$ in the spanwise direction z. Therefore, one must choose $K_x \geq 4$, $K_z \geq 2$; since a larger domain size improve statistics, K_x and K_z should exceed these minimum values by a generous margin.

As a velocity scale, it is appropriate to introduce the *friction velocity* $u_\tau = (\tau_w/\rho)^{1/2}$, in which ρ is the fluid's density and τ_w is the absolute value of the mean wall shear stress, related in its turn to the absolute value of the longitudinal pressure gradient $\partial p/\partial x$ by $\tau_w = \delta |\partial p/\partial x|$. The friction velocity provides the order of

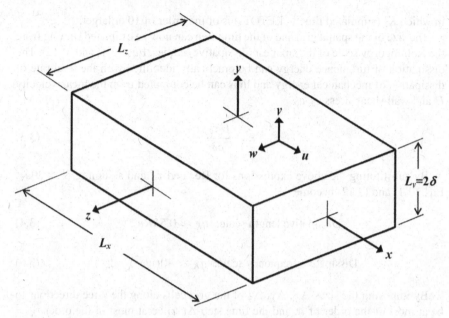

Fig. 3.1 Computational domain for the simulation of turbulent plane Poiseuille flow

magnitude of the turbulent velocity fluctuations. With the above definitions, the friction velocity u_τ, the mean velocity U (averaged both in time and over the channel's cross section) and the Fanning friction coefficient C_f are linked by the relation $u_\tau = (C_f/2)^{1/2}U$. For a typical value of C_f of ~ 0.005 one has $u_\tau \approx U/20$.

Similarly, besides the *bulk Reynolds number* $Re = UD_{eq}/\nu$ (based on mean velocity and hydraulic diameter, which is 4δ in an indefinite plane channel), it is useful to introduce the *friction velocity Reynolds number* $Re_\tau = u_\tau\delta/\nu$ (based on friction velocity and channel half height). For $C_f \approx 0.005$, from the definitions one has $Re_\tau \approx Re/80$.

The total duration t_{TOT} of the simulation must represent a statistically steady state, and thus must be much larger than the largest time constant expected in the time-dependent behaviour of the various flow quantities. Such a time constant is the *Large Eddy TurnOver Time* (LETOT), which corresponds to the mean lifetime of the largest turbulent structures (eddies). For a plane channel, it can be assumed equal to δ/u_τ so that, on the basis of the above definitions, it can be computed as

$$1 \text{ LETOT} = (\delta^2/\nu)Re_\tau^{-1} \tag{3.1}$$

Therefore, t_{TOT} can be estimated as

$$t_{TOT} \approx K_t(\delta^2/\nu)Re_\tau^{-1} \tag{3.2}$$

in which K_t (simulated time in LETOTs) is of the order of 10 or larger.

The size of the spatial grid and of the time step can now be estimated starting from the Kolmogorov scale of the smallest (dissipative) eddies, Eqs. (2.1) and (2.12). The dissipation of turbulence energy can be practically identified with the total rate of dissipation of mechanical energy and thus can be computed from the mean velocity U and wall shear stress τ_w as

$$\varepsilon = \frac{U\tau_w}{\rho\delta} \tag{3.3}$$

By substituting the above expressions for Re_τ and τ_w and assuming $U \approx 20u_\tau$, Eqs. (2.1) and (2.12) become

$$\text{Dissipative length scale: } \eta_K \approx 0.5\delta Re_\tau^{-3/4} \tag{3.4}$$

$$\text{Dissipative frequency scale: } f_K \approx 40(\nu/\delta^2) Re_\tau^{7/4} \tag{3.5}$$

By imposing the sizes Δx, Δy, Δz of the grid cells along the three directions to be at most of the order of η, and the time step Δt to be at most of the order $t_K = 1/f_K$, one has

$$\Delta x, \Delta y, \Delta z \leq 0.5\delta Re_\tau^{-3/4} \tag{3.6}$$

$$\Delta t \leq 0.025(\delta^2/\nu)Re_\tau^{-7/4} \tag{3.7}$$

Equations (3.6)–(3.7) imply the Courant criterion $\Delta t < \Delta x/U$ (Roache 1972), which imposes a *stability* limit if an explicit time stepping method is used, but remains an *accuracy* criterion also if unconditionally stable implicit schemes are adopted.

From Eq. (3.6) one obtains the minimum number of grid cells:

$$N_p = (L_x/Dx)\left(L_y/Dy\right)(L_z/Dz) \geq 16K_x K_z Re_\tau^{9/4} \tag{3.8}$$

while Eqs. (3.2) and (3.7) yield the minimum number of time steps:

$$N_t = t_{TOT}/\Delta t \geq 40K_t Re_\tau^{3/4} \tag{3.9}$$

A more accurate estimate of the minimum values of N_p and N_t could be conducted by accounting for the variation of C_f with Re and for the non-uniformity of turbulence in the near-wall regions (Grötzbach 1986; Zuniga Zamalloa 2012), but results would not change much. Equations (3.8)–(3.9) show how the computational effort increases as the friction velocity Reynolds number increases. Assuming the CPU time per time step to increase linearly with the number of grid points, the total CPU time will be proportional to $N_p N_t$ and thus to Re_τ^3.

As a more quantitative example, assume $K_x = 8$, $K_z = 4$, $K_t = 20$. Equations (3.8) and (3.9) yield $N_p \geq 512 \, \mathrm{Re}_\tau^{9/4}$, $N_t \geq 800 \, \mathrm{Re}_\tau^{3/4}$. If $\mathrm{Re}_\tau = 100$, corresponding to $\mathrm{Re} \approx 8000$ (weakly turbulent flow), one needs at least 16 million grid points and 25,000 time steps, a computational load within the reach of a contemporary (2020) multicore PC (although computing times will probably be of several days). However, if $\mathrm{Re}_\tau = 1000$, corresponding to $\mathrm{Re} \approx 80,000$ (fully turbulent flow), N_p and N_t increase 178-fold and 5.6-fold, respectively (to at least 2.8 *billion* cells and 140,000 time steps), making the simulation accessible only to powerful clusters with large-scale parallelism.

Taking into account what has been said in Sect. 2.5, it is clear that the DNS of the turbulent transport of *scalars* (e.g. specific enthalpy or concentration) with $\mathrm{Sc} > 1$ is more demanding than a purely hydrodynamic simulation, and may easily become prohibitive. In fact, in this case the grid must not resolve the Kolmogorov length scale η_K of dissipative eddies, but the smaller Batchelor scale $\eta_B = \eta_K/\mathrm{Sc}^{1/2}$. For $\mathrm{Sc} \approx 10^3$, typical of many solutes, $\eta_B \approx \eta_K/30$; for developed turbulence, no realistic simulation can be run with this spatial resolution even on the most powerful existing clusters.

3.3 DNS and Computing Performances

In order to make the above estimates more quantitative, one may assume that, provided numerical methods optimized for DNS are used, the necessary number of floating point operations is of the order of 10^4 per grid cell and per time step. In this case, the above estimates of N_p and N_t yield a minimum total number of floating point operations $N_{FLOP} = 10^4 N_p N_t \approx 4 \times 10^9 Re_\tau^3$.

Let $S_{\max}$ be the maximum theoric computational speed, in number of floating point operations per second (FLOPS), and $\chi \, S_{\max}$ be its actual speed in CFD calculations (with a computational efficiency χ typically ranging from 5 to 20%). The *throughput time tt* in seconds will be

$$tt \approx 4 \times 10^9 Re_t^3/(\chi \, S_{\max}) \tag{3.10}$$

One may also assume that the minimum number of variables to keep in memory is of the order of 25 per grid cell. Therefore, assuming double precision is used with 8 bytes per variable, the required memory in bytes will be

$$\mathrm{RAM} \approx 10^5 Re_{\tau s}^{9/4} \tag{3.11}$$

Currently (2020), for a common multi-core PC one may assume $\chi \, S_{\max} \approx 10^{10}$ (10 GFLOPS) so that, in the above example with $Re_\tau = 100$, one has $tt \approx 4 \times 10^5$ s (4–5 days), $\mathrm{RAM} \approx 3.2 \times 10^9$ bytes (3.2 Gbytes). Therefore, the more strict between throughput time and available RAM memory constraints is that relative to time.

For the largest existing clusters one may assume $\chi S_{max} = 2 \times 10^{16}$ (20 PFLOPS) so that, in the example with $Re_\tau = 1000$, one has $tt \approx 2 \times 10^2$ s (a bit more than 3 min) and RAM $\approx 5.6 \times 10^{14}$ bytes (560 Tbytes). Therefore, in this case it is RAM memory that imposes the stricter constraint.

Figure 3.2 summarizes the behaviour of N_p, N_t, RAM and tt, estimated by the above formulae, as functions of the friction velocity Reynolds number Re_τ under three different assumptions for the actual calculation speed: 1 GFLOPS, 1 TFLOPS and 1 PFLOPS.

To asses the past, present and future viability of DNS, the above estimates should be compared with the performance of computers. Figure 3.3 reports, from 1985 to 2020, the top computational performances worldwide: number of cores, computational speed per core in FLOPS and overall theoric computational speed (S_T, still expressed in FLOPS). As mentioned above, the actual computational speed depends on the problem studied (in particular, on the feasible degree of parallelization) and, for CFD applications, can be as low as 5% of the top speed.

The graph shows that, starting from the 1990s, the rate of increase of the computational speed per single processor has markedly slowed down. However, about at the same time there has been an increase in the size of parallel clusters of processors (often provided with multiple cores), from the few hundreds of cores of Fujitsu's *Numerical Wind Tunnel* (years 1990) to the several thousands of the ASCI series (around the year 2000) up to the several millions of the most powerful current super-clusters (for example, the Japanese cluster *Fugaku*, which currently—2020—detains the record for peak speed, includes 7,299,032 cores).

As a combined effect of the (weak) increase in computational speed per core and of the (fast) increase in the number of parallel cores, peak performances have continued to increase from the 1990s to now with a trend similar to that observed during the 1970s and 1980s, characterized by a performance doubling every ~ 14 months (this trend is often called *Moore's law*, but Moore actually addressed the density of transistors on a chip, only indirectly connected with computational speed).

Thus, the GFLOPS threshold (10^9 floating points operations per second) was first attained in 1988 (by *Cray-2*), the TFLOPS one (10^{12} FLOPS) in 1997 (by Intel's *ASCI Red*), the PFLOPS one (10^{15} FLOPS) in 2008 (by DOE's *Roadrunner*) and the EFLOPS (10^{18} FLOPS) threshold is now at hand, since *Fugaku* has achieved in 2020 a peak speed of 514 PFLOPS. At the same time, also the overall RAM memory has increased (although more slowly), from the 2 Gbytes of Cray-2 to the 4.86 Pbytes of *Fugaku*.

Therefore, the frontiers of the feasible in DNS (and in CFD in general) are rapidly moving forward. A DNS at $Re_\tau = 10^3$ implies some 4×10^{18} floating point operations; on the fastest existing computers of the time it required throughput times of a thousand years in 1988 (1 GFLOPS) and one year in 1998 (1 TFLOPS), being practically prohibitive up to that date. Such a simulation has actually been performed by Del Alamo et al. (2004) (30 TFLOPS, or about ten days throughput time), and could now be performed in about two minutes (at least in principle!) on the fastest clusters.

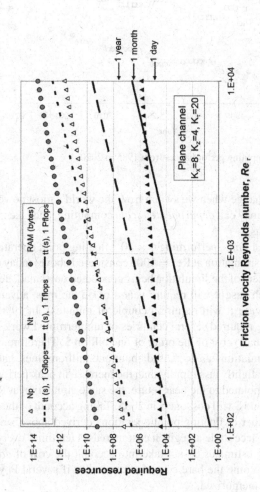

Fig. 3.2 Computational resources needed for the DNS of turbulent flow in a plane channel as functions of the friction velocity Reynolds number

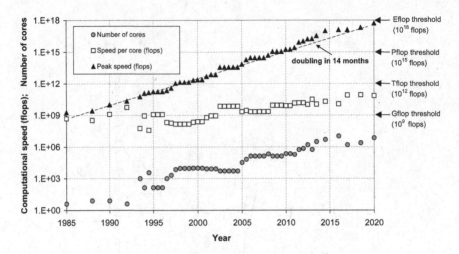

Fig. 3.3 Evolution of computing performances from 1985 to 2020

Of course, things change when we switch from the world's most powerful super-computers to what we may call *departmental-scale* computing, and take into account also *cost* considerations.

Using dollars as cost units, performing 4×10^{18} floating point operations (corre-sponding to the whole simulation at $Re_\tau = 1000$ considered above) today costs about a thousand \$, independent of the simulation being conducted on a small, departmental scale, cluster (in which case it will require years) or on the most advanced super-computers (in which case it will require a couple of minutes). In 1990, the same simulation would have required several centuries of uninterrupted work for a Cray-2, equivalent to a (virtual!) cost of the order of one billion \$. Therefore, in 30 years the cost of a given simulation has decreased about one million times, halving every ~ 18 months, i.e. only slightly less rapidly than the increase in peak performances. If this trend can be extrapolated to the near future, the same simulation at $Re_\tau = 1000$ that costs now a thousand \$ will cost one \$ in 2035. Taking account of the correlation between Re_τ and number of floating point operations derived above, one may also say that the Re_τ value accessible at a given cost increases ten times every 15 years.

Note that the above estimates do not take into account the cost of *data storage*, which may easily overcome the bare cost of computation if several Pbytes of data are to be permanently memorized.

3.4 An Example of Very High-Resolution DNS

As an impressive example of DNS, consider the simulations conducted by Lee et al. (2013) at the University of Texas. The authors simulated the turbulent flow of a constant-property fluid in a plane channel at $Re_\tau = 5200$ with periodic streamwise and spanwise conditions, as in Fig. 3.1. They used spectral methods (expansion in Fourier series along the homogeneous x and z directions and 7th-order basis-spline functions along y). The dimensions of the computational domain were $L_x = 24\delta$, $L_z = 16\delta$ and, of course, $L_y = 2\delta$ (by the present nomenclature, $K_x = 24$, $K_y = 2$, $K_z = 16$).

Following preliminary simulations with both coarser and finer grids, the final simulation was conducted using 10,240 Fourier modes along x, 7680 Fourier modes along z and 1536 grid points along y. For reasons that the specialists of spectral methods call *de-aliasing*, the number of collocation points along the homogeneous x and z directions must be 3/2 times the number of resolved Fourier modes, so that the spatial resolution of the simulation corresponds to a grid with $N_x = 15,360$, $N_y = 1536$ and $N_z = 11,520$, for a total ($N_p = N_x N_y N_z$) of about 240 *billion* Degrees Of Freedom (DOF). This number exceeds by ~ 15 times that of the most resolved previous DNS work, performed by Hoyas and Jiménez (2006) for $Re_\tau \approx 2000$. The simulation was protracted for 13 flow-through times, each resolved by 50,000 time steps, for a total of 650,000 steps.

Strictly speaking, the spatial resolution adopted does not meet the criteria discussed in Sect. 3.2: in fact, for $Re_\tau = 5200$, $K_x = 24$ and $K_z = 16$, Eq. (3.8) yields $N_p \geq 1.4 \times 10^{12}$, about 6 times larger than the adopted 240×10^9 *DOF*. Also the duration of the simulation is less than the 20 LETOTs usually regarded as appropriate for the statistical convergence of the results; this is partly compensated by the large size of the computational domain.

The simulation ran on a subset of 524,288 cores among the overall 760,000 of the supercomputer *Mira* (Argonne National Laboratories, Chicago), offering a peak speed of ~ 10 PFLOPS. The actual computational speed was much lower, of the order of 270 TFLOPS, with an efficiency of less than 3%. The overall CPU time was ~ 260 million core-hours, so that the throughput time (*tt*) was ~ 500 h (~3 s per time step). In economical terms, estimating the cost of *Mira* to be ~ 10,000 \$/h, the simulation cost some 5 million \$ (of course, CPU time was not actually purchased but rather ensured by research grants).

The scientific purpose of this piece of research was to clarify the dynamics of turbulence in the buffer region existing between the viscous sublayer and the outer flow, a region that acquires a significant spatial extension and a significant relevance only for sufficiently high Reynolds numbers. However, this simulation was also a rather spectacular demonstration of the power of supercomputing as applied to problems like hydrodynamic turbulence.

For a recent review of the nexus between computing power and insight into turbulence, see Jiménez (2020).

References

Comte-Bellot G (1963) Contribution a l'étude de la turbulence de conduit. Dissertation, Université de Grenoble, France

Del Alamo JC, Jiménez J, Zandonade P, Moser RD (2004) Scaling of the energy spectra of turbulent channels. J Fluid Mech 500:135–144

Grötzbach G (1986) Direct numerical and large eddy simulation of turbulent channel flows. In: Cheremisinoff NP (ed) Encyclopedia of fluid mechanics, vol 6. Gulf Publishing Co

Hoyas S, Jimenez J (2006) Scaling of the velocity fluctuations in turbulent channels up to $Re_\tau = 2003$. Phys Fluids 18:011702

Jiménez J (2020) Computers and turbulence. Eur J Mech B/fluids 79:1–11

Lee M, Malaya N, Moser RD (2013) Petascale Direct Numerical simulation of turbulent channel flow on up to 786k cores. In: Proceedings SC13—international conference for high performance computing, networking, storage and analysis, Denver, Colorado, November 17–21, Article No. 61. ACM publishers, New York

Roache P (1972) Computational fluid dynamics. Hermosa Publishers, Albuquerque, NM

Zuniga Zamalloa CC (2012) Experiments on turbulent flows in rough pipes: spectral scaling laws and the spectral link. PhD Dissertation in Theoretical and Applied Mechanics, University of Illinois at Urbana-Champaign, USA

Chapter 4
Large Eddy Simulation

*Turbulence is the gateway through which large fluid masses in
ordered motion march to their heat-death doom*
Carl F. von Weiszäcker

Abstract *Large Eddy Simulation* (LES) is a turbulence modelling technique based
on spatial filtering, which leaves all quantities still functions of time. The rationale
for LES is discussed, different possible filters are presented, the unresolved (*subgrid*)
terms arising from filtering are derived, and alternative subgrid models are reviewed
and compared, including the classic Smagorinsky model and the more advanced
"dynamic" one.

Keywords Turbulence modelling · Large eddy simulation · Spatial filter · Subgrid
viscosity · Smagorinsky model · Dynamic model

4.1 Spatial Filtering

Large Eddy Simulation (LES) is an approach to turbulence modelling based on
the *spatial filtering* of the flow field. The rationale for this approach is that, in a
turbulent flow, the large-scale structures (*large eddies*), which contain most of the
mechanical energy of turbulence and account for most of the advective momentum
and scalar transport, do not lend themselves to be described by a universal model
because they are highly anisotropic and problem-dependent; therefore, they are better
explicitly simulated. On the other hand, small-scale structures (close to the dissipative
Kolmogorov scale) are almost isotropic and thus problem-independent, contain only
a small fraction of the turbulence energy, and thus are more amenable to being
described by simple models.

In LES, the first conceptual step is to define a spatial filter function $G(\mathbf{x}, \mathbf{y})$,
formally a function of the six coordinates of points $\mathbf{x}$ and $\mathbf{y}$ but often a function only
of the distance $|\mathbf{x} - \mathbf{y}|$; the application of this filter to the generic quantity $\varphi(\mathbf{x}, t)$
gives rise to the decomposition of φ into a *filtered*, or *resolved*, component

M. Ciofalo, *Thermofluid Dynamics of Turbulent Flows*, UNIPA Springer Series,
https://doi.org/10.1007/978-3-030-81078-8_4

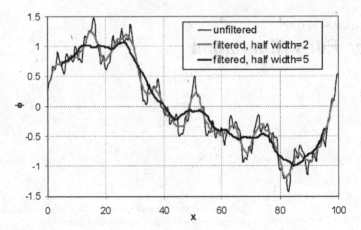

Fig. 4.1 Effect of spatial filters G of different half-widths on a generic quantity φ

$$\langle \varphi(\mathbf{x}, t) \rangle = \int G(\mathbf{x}, \mathbf{y}) \varphi(\mathbf{y}, t) d^3 \mathbf{y} \qquad (4.1)$$

and a *residual*, o *unresolved*, component

$$\varphi'(\mathbf{x}, t) = \varphi(\mathbf{x}, t) - \langle \varphi(\mathbf{x}, t) \rangle \qquad (4.2)$$

It is convenient for the filter to be a *linear operator*.

An example of the application of a filter to a generic quantity φ (supposed, for the sake of simplicity, to be a function of only one spatial variable x) is reported in Fig. 4.1. Here, the unfiltered quantity is defined in $(0, 100)$ and contains wavelengths Λ from 0.5 to 50 (i.e., wavenumbers n from 0.02 to 2). The filter is a simple running average in $(x - \Delta/2, x + \Delta/2)$; its effect is shown for two different values of the half-width $\Delta/2$, namely, 2 and 5.

The application of the filter cancels all details with wavelength $< \Delta$, while preserving the large-scale behaviour of the function φ. It must be stressed that the filtered function $\langle \varphi \rangle$ is still a continuous function of x, defined in the whole domain of existence of φ (with some special treatment required in the boundary regions).

Different filter functions G have been proposed and applied in LES. It is instructive to consider their effect both in the physical, or configuration, space of the coordinates $\mathbf{x}$ and in the corresponding transformed, or Fourier, space of wavenumbers $\mathbf{n}$. For the one-dimensional case, three alternative filter functions G are schematically shown in Fig. 4.2; in three dimensions, their expressions are:

- *box*, or *top-hat*, filter, defined in physical space $\mathbf{x}$ as:

$$G(\mathbf{x}, \mathbf{y}) = 1/\Delta^3 \quad \text{if } |x_i - y_i| \le \Delta/2 \ (i = 1, 2, 3) \qquad (4.3)$$

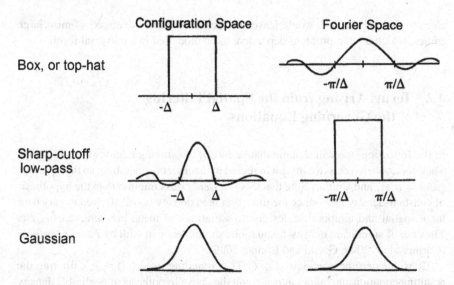

Fig. 4.2 Alternative spatial filters shown in physical, or configuration, space (left) and in transformed, or Fourier, space (right)

$$G(\mathbf{x}, \mathbf{y}) = 0 \quad \text{otherwise} \tag{4.4}$$

- *sharp cutoff* low pass filter, defined in Fourier space **n** as:

$$G(\mathbf{n}) = 1 \quad if \ |n_i| < \pi/\Delta . \ (i = 1, 2, 3) \tag{4.5}$$

$$G(\mathbf{n}) = 0 \quad \text{otherwise} \tag{4.6}$$

- *Gaussian* filter, defined in physical space **x** as:

$$G(\mathbf{x}, \mathbf{y}) = \left[\frac{1}{\sqrt{\pi} \cdot \Delta}\right]^3 \exp\left[-\frac{1}{\Delta^3}(\mathbf{x} - \mathbf{y})^2\right] \tag{4.7}$$

and in Fourier space **n** as:

$$G(\mathbf{n}) = \exp\left(-\frac{\Delta^2 n^2}{4}\right) \tag{4.8}$$

The Gaussian filter has the same shape in both spaces, while box and sharp-cutoff filters can be regarded as dual filters with respect to physical and transformed space.

With reference to the typical spectrum schematically shown in Fig. 2.3, the width Δ of the spatial filter should be chosen so that $1/\Delta$ falls within the wavenumber interval representing the inertial sub-range of turbulence. Values of Δ lower than the Kolmogorov scale η_K would turn LES into DNS, while values excessively large and close to the large scale L would leave as unresolved terms the effects of most large eddies, which are too problem-dependent to be modelled in a universal form.

4.2   Terms Arising from the Spatial Filtering
of the Governing Equations

In the following, we will assume that *turbulent density fluctuations are negligible*. Therefore, ρ behaves as a constant in regard to the filtering operation, so that $\langle \rho \rangle = \rho$, $\langle \rho u_i \rangle = \rho \langle u_i \rangle$ and similar. Note that this is a *weaker* assumption than the hypothesis of constant-density flow, since the time-averaged density is still allowed to vary over large spatial and temporal scales due to variations of mean pressure, density etc. The case of significant density fluctuations should be dealt with by *Favre averaging* (Garnier et al. 2009; Gatski and Bonnet 2013).

Consider first the continuity Eq. (1.7). Expanding $u_i = \langle u_i \rangle + u_i'$, filtering the resulting equation and taking into account the above hypothesis of negligible density fluctuations, it reduces to

$$\frac{\partial \rho}{\partial t} + \frac{\partial \rho \langle u_j \rangle}{\partial x_j} = 0 \qquad (4.9)$$

which states that the filtered velocity field satisfies the same continuity equation as the unfiltered one. For constant-density fluids Eq. (4.9) reduces to $\partial \langle u_j \rangle / \partial x_j = 0$, which states that the filtered velocity field is, like the primitive one, divergence-free.

Consider now the generic momentum (Navier–Stokes) Eq. (1.17), repeated here below for the benefit of the reader

$$\frac{\partial \rho u_i}{\partial t} + \frac{\partial \rho u_j u_i}{\partial x_j} = -\frac{\partial p^*}{\partial x_i} + \frac{\partial}{\partial x_j}\left[\mu \left(\frac{\partial u_i}{\partial x_j} + \frac{\partial u_j}{\partial x_i} \right) \right] + \rho a_i \qquad (1.17 \text{ rept.})$$

in which $p^* = p + 2/3 \mu S_{kk} = p + 2/3 \mu \nabla \mathbf{u}$.

Substitute $\langle u_i \rangle + u_i'$ for each velocity u_i and $\langle p^* \rangle + p^{*\prime}$ for the modified pressure p^*, develop the products of sums and then filter the resulting equation term by term. Remembering that filter and partial derivative *commute* and assuming turbulent fluctuations of ρ and $s\mu$ to be negligible, the result is

$$\frac{\partial \rho \langle u_i \rangle}{\partial t} + \frac{\partial \rho \langle u_i \rangle \langle u_j \rangle}{\partial x_j} = -\frac{\partial \langle p^* \rangle}{\partial x_i} + \frac{\partial}{\partial x_j}\left[\mu \left(\frac{\partial \langle u_i \rangle}{\partial x_j} + \frac{\partial \langle u_j \rangle}{\partial x_i} \right) \right] + \frac{\partial \wp_{sg,ij}}{\partial x_j} + \rho \langle a_i \rangle$$

$$(4.10)$$

i.e., equations formally similar to the primitive Navier–Stokes ones (1.17), except:

- the unknowns are not primitive velocities and pressure, but their *filtered* counterparts;
- the RHS now includes gradients of the additional terms $\wp_{sg,ij}$ (*subgrid stresses*), similar to the stresses $\wp_{ij}$ but arising from filtering the nonlinear terms in Eq. (1.17).

It is easy to show, by carefully performing in detail the passages which are only sketched above, that the additional terms $\wp_{sg,ij}$ can be written as

$$\wp_{sg,ij} = L_{ij} + C_{ij} + R_{ij} \tag{4.11}$$

in which the three components L_{ij}, C_{ij}, R_{ij} are

- *Leonard terms* $L_{ij} = -\rho\big(\langle u_i\rangle\langle u_j\rangle\big) - \langle u_i\rangle\langle u_j\rangle$ (4.12)

- *cross terms* $C_{ij} = -\rho\big(\langle u_i\rangle u_j'\big) + \langle u_i'\langle u_j\rangle\big)$ (4.13)

- *unresolved terms* $Rij = -\rho\langle u_i' u_j'\rangle$ (4.14)

Now, Leonard terms contain only filtered (resolved) quantities and do not require any modelling of unresolved quantities. Cross terms contain a combination of resolved and unresolved quantities and require some modelling; Leonard (1974) showed that, if the Gaussian filter of width Δ is used, Leonard and cross terms can be approximated as

$$L_{ij} \approx -\rho(\Delta^2/24)\nabla\big(\langle u_i\rangle\langle u_j\rangle\big) \tag{4.15}$$

$$C_{ij} \approx -\rho(\Delta^2/24)\big(\langle u_i\rangle\nabla\langle u_j\rangle + \langle u_j\rangle\nabla\langle u_i\rangle\big) \tag{4.16}$$

The terms R_{ij}, on the other hand, are *unresolved stresses* proper, corresponding to the Reynolds stresses arising in RANS turbulence models (see Chap. 5), and there is no way of approximating them in terms of resolved quantities on purely mathematical grounds. Rather, some physical assumptions have to be made and translated into what is commonly called a *subgrid model*.

The modelling of Leonard, cross and unresolved terms raises subtle problems of consistence and invariance of the filtered equations and has been the subject of a vast literature (Speziale 1985; Germano 1992). In order to avoid these complications, some authors prefer globally to model the sum of all additional terms, $\wp_{sg,ij} = L_{ij} + C_{ij} + R_{ij}$, or at least the sum of the true unresolved terms, $C_{ij} + R_{ij}$, rather that the unresolved stresses R_{ij} alone.

In the context of finite-volume computational methods, a natural choice of filtering is averaging over each finite volume, which is necessary anyway for the discretization of the problem (indpendent of turbulence). In this case, Leonard and cross terms *vanish* and the additional terms reduce themselves to the unresolved stresses R_{ij}, which can legitimately be called *sub-grid*. The price to pay is that the filtered functions are no more continuous functions over the computational domain, and are only defined at the centroids (or other specific points) of the finite volumes. Moreover, identifying the spatial filter with the average over each finite volume causes the filter itself to become non-uniform when selectively refined or body-fitted grids are adopted; in its turn, this gives rise, at least in principle, to extra terms in the filtered equations (Schumann 1975).

Despite these shortcomings, in the following, having mainly in mind numerical solutions by the finite volume method, we will identify the subgrid stresses in Eq. (4.10) with the unresolved stresses $R_{ij} = -\rho\langle u_i{'}u_j{'}\rangle$ and will not deal any further with Leonard and cross terms.

Manipulations similar to those performed for the viscous stresses in the Appendix Sect. A.4.2, and in Sect. 1.3, Eq. (1.15), lead to expressing the unresolved stress tensor as the sum of an isotropic compressive stress (*subgrid pressure* p_{sg}) and a traceless part:

$$R_{ij} = -p_{sg}\delta_{ij} + (R_{ij} + p_{sg}\delta_{ij}) = -p_{sg}\delta_{ij} + \tau_{sg,ij} \qquad (4.17)$$

in which (in explicit form) $p_{sg} = -R_{kk}/3 = \rho\langle u{'}_k u{'}_k\rangle/3$.

By defining a *subgrid kinetic energy* k_{sg} as

$$k_{sg} = \langle u{'}_k u{'}_k\rangle/2 = (\langle u'^2\rangle + \langle v'^2\rangle + \langle w'^2\rangle)/2 \qquad (4.18)$$

the subgrid pressure can be written as $p_{sg} = (2/3)\rho k_{sg}$ and thus the decomposition in Eq. (4.17) becomes $R_{ij} = -(2/3)\rho k_{sg}\delta_{ij} + \tau_{sg,ij}$. By using this last expression for R_{ij} (identified with $\wp_{sg,ij}$ for the reasons discussed above) and substituting for $\wp_{sg,ij}$ into Eq. (4.10), one obtains:

$$\frac{\partial\rho\langle u_i\rangle}{\partial t} + \frac{\partial\rho\langle u_i\rangle\langle u_j\rangle}{\partial x_j} = -\frac{\partial p^{**}}{\partial x_i} + \frac{\partial}{\partial x_j}\left[\mu\left(\frac{\partial\langle u_i\rangle}{\partial x_j} + \frac{\partial\langle u_j\rangle}{\partial x_i}\right)\right] + \frac{\partial\tau_{sg,ij}}{\partial x_j} + \rho\langle a_i\rangle$$

$$(4.19)$$

in which $p^{**} = \langle p^*\rangle + (2/3)\rho k_{sg}$ is a *second modified pressure* including not only the divergence term $(2/3)\mu\nabla\mathbf{u}$ but also the subgrid energy term $(2/3)\rho k_{sg}$.

In regard to scalar transport, and, in particular, heat transport, manipulations similar to those described for the Navier–Stokes Eq. (1.17) can be applied to Eq. (1.28), yielding (for negligible fluctuations of λ and c_p):

$$\frac{\partial\rho c_p\langle T\rangle}{\partial t} + \frac{\partial\rho c_p\langle u_j\rangle\langle T\rangle}{\partial x_j} = \frac{\partial}{\partial x_j}\left(\lambda\frac{\partial\langle T\rangle}{\partial x_j}\right) + q''' - \frac{\partial}{\partial x_j}q_{sg,j} \qquad (4.20)$$

in which the terms $q_{sg,j}$ represent additional (subgrid) heat fluxes which, apart from mixed contributions similar to L_{ij} and C_{ij}, which vanish in filtering by finite volume averaging, consist of cross-correlations of unresolved velocities and temperatures, $\rho c_p \langle u_j{}' T' \rangle$.

An important observation is that, albeit additional stresses and fluxes are generally called, by a consolidated tradition, *subgrid terms*, yet this denomination is inaccurate, and a more rigorous expression would be *subfilter terms*. In fact, they are terms arising from the filtering step, before any discretization of the computational domain by a grid. Actually, the computational grid could be, and should be, finer than the filter, while the opposite would be inconsistent. As was mentioned above, filter size and grid size coincide if implicit filtering by volume averaging is adopted.

4.3 The Smagorinsky Sub-Grid Model

Additional stresses and fluxes have now to be modelled in order to achieve closure of the system of filtered Eqs. (4.9), (4.19) and, if appropriate, (4.20).

A vast class of models is based on the Bousinnesq-like hypothesis that the traceless part of the subgrid stress tensor be proportional to the traceless part of the resolved strain rate tensor:

$$\tau_{sg.ij} = 2\mu_{sg} \left[\langle S_{ij} \rangle - \frac{1}{3}\delta_{ij}\langle S_{kk} \rangle \right] \tag{4.21}$$

μ_{sg} being a *sub-grid viscosity* which will be, in general, a function of $\langle S_{ij} \rangle$ or other flow variables, according to a law to be specified. Of course, the trace of $\langle S_{ij} \rangle$ vanishes in constant-density flows, so that the subtraction of $\langle S_{kks} \rangle /3$ from the diagonal terms can be omitted.

Equation (4.21) is built by analogy with the similar Boussinesq assumption for viscous stresses, see Sect. A.4.2 of the Appendix, and is consistent with the heuristic model of turbulence sketched in Sect. 2.9. It eventually rests on the view that unresolved turbulent eddies undergo *diffusive* (Brownian) motions as, at a lower scale, the fluid's molecules. Models based on Eq. (4.21) are also called *gradient diffusion* models.

With manipulations similar to those used to derive the primitive Navier–Stokes Eq. (1.17), Eqs. (4.19) and (4.21) together yield the subgrid-viscosity form of the filtered Navier–Stokes equations to be solved in LES:

$$\frac{\partial \rho \langle u_i \rangle}{\partial t} + \frac{\partial \rho \langle u_i \rangle \langle u_j \rangle}{\partial x_j} = -\frac{\partial p^{**}}{\partial x_i} + \frac{\partial}{\partial x_j}\left[(\mu + \mu_{sg})\left(\frac{\partial \langle u_i \rangle}{\partial x_j} + \frac{\partial \langle u_j \rangle}{\partial x_i} \right) \right] + \rho \langle a_i \rangle$$

$$\tag{4.22}$$

These are similar to the primitive Navier–Stokes Eq. (1.17), except that velocities must be interpreted as filtered variables, pressure is modified as specified above, and the molecular viscosity μ is replaced by the total (i.e., molecular + subgrid) viscosity $\mu + \mu_{sg}$.

The problem of *closure* is now reconducted to that of expressing the subgrid viscosity as a function of resolved (filtered) fields. Dimensional analysis and heuristic arguments similar to those behind Eq. (2.22) show that the subgrid viscosity can be expressed (apart from a numerical coefficient) as the product of three terms: density, length scale squared, and strain rate. Among the models that can be built on this basis, the most widely used has been for long years that proposed by Smagorinsky (1963):

$$\mu_{sg} = \rho(c_S \Delta)^2 (2\langle S_{ij}\rangle\langle S_{ij}\rangle)^{1/2} \tag{4.23}$$

where Δ is the local width of the filter (coinciding with the local size of the computational grid if implicit filtering by volume averaging is adopted), c_S is a model constant of the order of 10^{-1} and the term $2\langle S_{ij}\rangle\langle S_{ij}\rangle$ is the quadratic invariant of the resolved strain rate tensor $\langle S_{ij}\rangle$. Note that $2\langle S_{ij}\rangle\langle S_{ij}\rangle$ is proportional, through the molecular viscosity μ, to the direct mechanical energy dissipation $2\mu\langle S_{ij}\rangle\langle S_{ij}\rangle$ and, through the subgrid viscosity μ_{sg}, to the production of subgrid kinetic energy $\tau_{ij}\langle S_{ij}\rangle$ $= \mu_{sg}\langle S_{ij}\rangle\langle S_{ij}\rangle$ (incompressible fluids).

The Smagorinsky model is formally a mixing length model, in which the role of Prandtl's mixing length is played by the filter size Δ. It can be theoretically justified by the assumption of local equilibrium between mechanical energy passing from resolved to subgrid scales and viscous dissipation at the lowest (dissipative) scales $\sim\eta_K$. A derivation of the model from the so called *Direct Interaction Theory* (DI) was given by Yoshizawa (1982). The value of c_S was theoretically obtained by Lilly (1966) as ~0.23 for isotropic turbulence; however, generally lower values (from 0.08 to 0.2) have been used in most applications to flows exhibiting resolved velocity gradients (shear flows).

In order to account for the attenuation of turbulence scales occurring in the proximity of solid walls, the Smagorinsky model is often modified by multiplying the spatial scale Δ in Eq. (4.20) by a *damping factor* f_μ, e.g. in the form proposed by Van Driest (1956), Eq. (2.23). The resulting expression for μ_{sg} is

$$\mu_{sg} = \rho(c_S f \mu \Delta)^2 (2\langle S_{ij}\rangle\langle S_{ij}\rangle)^{1/2} \tag{4.24}$$

In regard to heat transport, the problem is to model the subgrid fluxes $q_{sg,j}$ in Eq. (4.20). Consistently with the gradient-diffusion assumption made for the subgrid stresses, Eq. (4.21), one may write

$$q_{sg,j} = -c_p \Gamma_{sg} \frac{\partial \langle T\rangle}{\partial x_j} \tag{4.25}$$

in which $\langle T\rangle$ is the resolved, or filtered, temperature and Γ_{sg} is the subgrid thermal diffusivity. In its turn, Γ_{sg} can be assumed to be proportional to the subgrid viscosity:

$$\Gamma_{sg} = \frac{\mu_{sg}}{\sigma_{sg}} \tag{4.26}$$

in which the proportionality constant σ_{sg} is called *subgrid* Prandtl n*umber*. For this parameter, different values, from 0.25 to 0.85, have been proposed in the literature depending on the specific problem studied.

Despite its simplicity, the Smagorinsky model (complemented, in non-isothermal problems, by the hypothesis of proportionality between the subgrid diffusivities of heat and momentum) has given satisfactory results in applications ranging from simple geometries (atmospheric boundary layer, Poiseuille flow, isotropic turbulence) to more complex ones (corrugated plate heat exchangers, spacer-filled channels).

As an example, Fig. 4.3 reports computational results obtained by Breuer and Rodi (1994) for the turbulent flow in a duct of square cross section at $Re_\delta \approx 2200$ using the Smagorinsky model with a computational grid of $62 \times 41 \times 41$ ($x \times y \times z$) cells. The problem is identical to that studied by Gavrilakis (1992) using DNS with a much finer grid ($1000 \times 127 \times 127$ cells).

A result worth noting is that, although the model is a gradient-diffusion one and thus does not separately provides the three (subgrid) normal stresses, the characteristic secondary flow, which arises from the asymmetry between the three (total) turbulent normal stresses, is correctly predicted: this happens because the turbulent structures down to the scale of the grid are explicitly simulated. Therefore, only *unresolved* (subgrid) normal turbulent stresses are modelled as isotropic, but *resolved* and thus *total* ones are not.

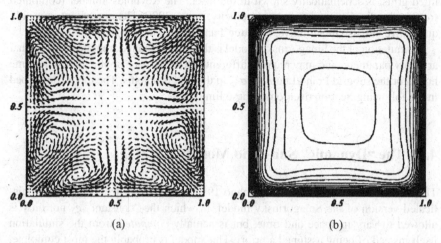

Fig. 4.3 LES results for fully developed turbulent flow in a square channel at $Re_\delta \approx 2200$ using $62 \times 41 \times 41$ ($x \times y \times z$) grid cells. **a** Vector plot of the secondary flow in the cross section yz; **b** contours of the streamwise velocity in the same plane. Reprinted from Breuer and Rodi (1994) with permission of Springer Nature

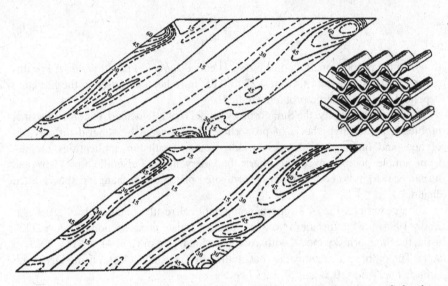

Fig. 4.4 Distribution of the Nusselt number on the walls of a unit cell of a corrugated-plate heat exchanger, whose geometry is shown in the inset. Top: experimental results from liquid crystal thermography; bottom: LES predictions using the Smagorinsky model with $c_s = 0.2$ and $\sigma_{sg} = 0.5$. From results in Ciofalo et al. (1996)

As a further example, Fig. 4.4 (Ciofalo et al. 1996) compares the experimental distribution of the local Nusselt number on the walls of a unit cell of a corrugated-plate heat exchanger (obtained by thermochromic liquid crystals and digital image processing) with LES predictions. The problem's geometry, such as to require body-fitted grids, is schematically shown in the inset. The Reynolds number (computed from hydraulic diameter and mean velocity) was ~3900. The agreement, not only qualitative but also quantitative, is better than that obtained by RANS models.

A limitation of the Smagorinsky model is that the constant c_S remains a somewhat arbitrary parameter and may require different values for each application. The same holds for the subgrid Prandtl number σ_{sg} in thermal problems. The model described in the following Section overcomes these limitations.

4.4   The "Dynamic" Sub-Grid Model

The "dynamic" subgrid model (Germano 1992) can be regarded as a more sophisticated version of the Smagorinsky model, in which the "constant" c_S not only is allowed to vary in space and time, but is actually *computed* from the simulation itself instead of being assigned a priori. This model is probably the most complete, consistent and flexible among those proposed; therefore, a fairly detailed description of it will be given in the following, despite the rather complex formalism involved.

The rationale for the dynamic model can be traced back to Bardina et al. (1980); it consists of sampling the smallest resolved scales and using this information to model the unresolved scales. To this purpose, besides the conventional LES filter $G(\varphi \rightarrow \langle \varphi \rangle)$ of width Δ_G (*grid filter*), a second filter $F(\varphi \rightarrow \{\varphi\})$ of larger width $\Delta_F = \zeta \Delta_G$ (test filter) is defined. Typically, $\zeta = 2$–3. Three distinct residual stresses can now be built:

$$\wp_{sg,ij} = -\rho\left(\langle u_i u_j \rangle - \langle u_i \rangle \langle u_j \rangle\right) \tag{4.27}$$

$$T_{ij}^* = -\rho\left(\{\langle u_i u_j \rangle\} - \{\langle u_i \rangle\}\{\langle u_j \rangle\}\right) \tag{4.28}$$

$$L_{ij}^* = -\rho\left(\{\langle u_i \rangle \langle u_j \rangle\} - \{\langle u_i \rangle\}\{\langle u_j \rangle\}\right) \tag{4.29}$$

- $\wp_{sg,ij}$ is the conventional *sub-grid stress*, which accounts for the unresolved scales below the grid filter G;
- T_{ij}^* is a *sub-test stress* which accounts for the scaled below the test filter F;
- L_{ij}^* is a *resolved turbulent stress*, which accounts for the scales intermediate between the two filters G and F; it is similar to the Leonard term, Eq. (4.12), but is built using two different filters G, F instead of a single filter G.

Between the three above quantities the following noteworthy identity holds, which can be derived in a pure formal way from the definitions:

$$L_{ij}^* = T_{ij}^* - \{\wp_{sg,ij}\} \tag{4.30}$$

Compute now $\wp_{sg,ij}$ and T_{ij}^* under the gradient-diffusion assumption (4.21), expressing the turbulent viscosity by the Smagorinsky model (4.23) and *the same c_S*:

$$\wp_{sg,ij} = 2\rho(c_S \Delta_G)^2 \langle S \rangle \langle S_{ij} \rangle \tag{4.31}$$

$$T_{ij}^* - \frac{1}{3}\delta_{ij} T_{kk}^* = 2\rho(C_S \Delta_F)^2 \{\langle S \rangle\}\{\langle S_{ij} \rangle\} \tag{4.32}$$

in which $\langle S \rangle = 2\langle S_{ij} \rangle \langle S_{ij} \rangle$. Substituting Eqs. (4.31) and (4.32) into Eq. (4.30) yields

$$L_{ij}^* = 2\rho(c_S \Delta G)^2 M_{ij}^* \tag{4.33}$$

in which the terms M_{ij}^* are defined as

$$M_{ij}^* = \zeta^2 \{\langle S \rangle\}\{\langle S_{ij} \rangle\} - \{\langle S \rangle \langle S_{ij} \rangle\} \tag{4.34}$$

Since both L_{ij}^* and M_{ij}^* contain only *resolved* quantities (i.e., quantities that are filtered at grid level, and thus explicitly predicted by the simulation), Eq. (4.30) allows, at least in principle, the Smagorinsky "constant" c_S to be computed, thus obtaining a subgrid model that is *completely free* from arbitrary parameters.

Actually, Eq. (4.33) *overdetermines* c_s (a different value may be computed for each of the six independent combinations of i and j), and is generally used in a *statistical* sense; following a proposal by Lilly (1992), both members are first tensorially contracted with M^*_{ij} and then averaged in space or time over a suitable *ensemble* (e.g. a homogeneous direction or plane or, if these are not available, a time interval). Denoting this kind of averaging by an overbar, one has

$$(c_S \Delta_G)^2 = \frac{1}{2\rho} \cdot \frac{\overline{L^*_{kl} M^*_{kl}}}{\overline{M^*_{ij} M^*_{ij}}} \qquad (4.35)$$

Subgrid *heat transfer* can be treated in a similar way. The resolved turbulent heat fluxes, corresponding to the Leonard terms L_{ij}, are (apart from the ρc_p factor)

$$Q^*_i = \{\langle u_i \rangle \langle T \rangle\} - \{\langle u_i \rangle\}\{\langle T \rangle\} \qquad (4.36)$$

Proceeding as for the stresses, the resolved turbulent heat fluxes are expressed as

$$Q^*_i = -\frac{(c_S \Delta_G)^2}{\sigma_{sg}} N^*_i \qquad (4.37)$$

in which the terms N^*_i, thermal equivalents of the terms M^*_{ij}, are given by

$$N^*_i = \zeta^2 \{\langle S \rangle\} \frac{\partial \{\langle T \rangle\}}{\partial x_i} - \left\{ \langle S \rangle \frac{\partial \langle T \rangle}{\partial x_i} \right\} \qquad (4.38)$$

Computing now the scalar product of both members in Eq. (4.37) by N^*_i and averaging as above, one obtains

$$\sigma_{sg} = -(c_S \Delta_G)^2 \frac{\overline{N^*_j N^*_j}}{\overline{Q^*_k N^*_k}} \qquad (4.39)$$

Equation (4.36) allows the subgrid Prandtl number to be *computed* at each spatial location and at each instant from resolved quantities, thus disposing of any arbitrary model constant.

As an example, Fig. 4.5 reports computational results by Fatica et al. (1994) for a turbulent round jet. The quantity represented is the instantaneous azimuthal average of the azimuthal vorticity ω_ϑ. Graph (a) reports DNS results obtained with a fine grid resolving all turbulence scales. Graph (b) reports the results of a coarse-grid simulation with no subgrid model (unresolved DNS); it exhibits numerical oscillations that viscous dissipation is not sufficient to damp. Finally, graphs (c) and (d) were obtained using the same coarse grid, but with the Smagorinsky and dynamic subgrid models, respectively. The *subgrid* dissipation is now sufficient to damp numerical

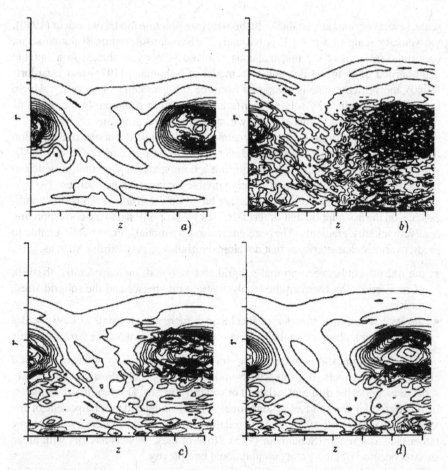

Fig. 4.5 Instantaneous azimuthal average of the azimuthal vorticity ω_ϕ in a round jet. **a** DNS, $129 \times 91 \times 129$ ($z \times r \times \phi$) cells; **b** DNS, $33 \times 65 \times 41$ cells; **c** LES-Smagorinsky, grid as in (**b**); **d** LES-dynamic model, grid as in (**b**). Reprinted from Fatica et al. (1994) with permission of Springer Nature

oscillations. The dynamic model predictions are slightly more regular and in better agreement with the DNS, but also the Smagorinsky model yields acceptable results, and the differences between the two subgrid models are rather small.

4.5 Other Sub-Grid Models

A family of subgrid models can be built by expressing the subgrid viscosity μ_{sg} as the product of density, length and velocity rather than of density, length squared and velocity gradient. The length scale is still the filter width Δ; in regard to the velocity scale, several options are available. In the *structure function* model of Lesieur (1990), the velocity scale in a point P is basically the standard deviation (valutata su un opportuno intorno) of the instantaneous resolved velocity, evaluated in a suitable neighborood of P. In the one-equation model of Schumann (1975) and Deardorff (1980), the velocity scale is computed from the *sub-grid kinetic energy* k_{sg}, which, in its turn, is obtained by solving a differential transport equation. In the past, this approach found applications especially in meteorological problems.

A combination of the Germano dynamic model and of the above one-equation model is the *Dynamic One Equation Model* (DOEM) proposed by Davidson (1997), which offers the advantage of not requiring a homogeneous direction in which to perform averages and thus is particularly suitable for complex geometries.

The subgrid models examined so far are based on the *gradient-diffusion* hypothesis, i.e. on the assumption that subgrid stresses are proportional to the corresponding resolved velocity gradients. They are *eddy viscosity* models, intrinsically unable to predict various characteristics that developed turbulence may exhibit, such as.

- the *anisotropy* between normal subgrid stresses, and, more generally, the lack of proportionality between the resolved strain rate tensor and the subgrid stress tensor;
- the *inverse spectral transfer* of mechanical energy from small to larger scales (*backscatter*), observed in the *wall region* and in phenomena like *vortex pairing*.

Note that similar shortcomings of the gradient-diffusion assumption are observed also in RANS models. Both in RANS and in LES, the only drastic solution is to separately model the different turbulent or subgrid stresses. This requires the solution of 6 differential transport equations, one for each independent component of the tensor $\tau_{sg,ij}$ (Deardorff 1974), or the modelling of the subgrid viscosity as a *tensor* rather than as a *scalar* (Schumann 1975). None of these models has met with much success because of their great computational complexity.

It must be remembered that, in LES, the *anisotropy* issue is much less relevant than in RANS models, since the small scales of turbulence do actually tend to isotropy. In regard to the *backscatter* issue, stochastic models have been proposed (Mason and Thomson 1992) which can take the phenomenon into account by slight modifications of Smagorinsky-type subgrid models.

4.6   Boundary and Initial Conditions for LES

LES may require a special attention to the correct description of boundary and initial conditions. Only a short account of these problems will be given here, see e.g. Ciofalo (1994) for a more detailed discussion.

First, on solid walls the no slip conditions can directly be applied only if the computational grid is fine enough to resolve the viscous sublayer (identified, in wall units ν/u_τ, by the condition $y^+ \leq \sim 11$), in which velocity profiles can be assumed linear. In high-Reynolds number problems, satisfying this condition may be difficult; a possible solution is to adopt *synthetic* wall boundary conditions based on "universal" *wall functions* (Piomelli et al. 1989), similar to those adopted in RANS models.

An alternative approach, known as *Detached Eddy Simulation*, or DES (Spalart et al. 1997), consists of using LES (intrinsically time-dependent) in the regions far from solid walls and RANS, possibly steady and based on simplified models such as one-equation ones, in the near-wall region. This approach allows the simulation of high-Reynolds number problems with relatively coarse grids. For applications and comparisons with "pure" RANS and LES results see e.g. Schmidt and Thiele (2002) or Viswanathan and Tafti (2006).

In *periodic* problems, such as the fully developed flow in ducts, periodicity conditions replace the explicit specification of *inflow* or *outflow* conditions. In all other cases, however, these conditions must be prescribed. *Outflow* conditions do not generally pose special difficulties with respect to RANS steady simulations; for example, the first or second derivatives of flow quantities along the direction normal to an outlet section can be prescribed. The consistent specification of time-dependent *inflow* conditions is more difficult; a common approach is to compute them by preliminary simulations of fully developed (axially periodic) flow in the *upstream* geometry. In some problems, the transport of turbulence into the computational domain from upstream is negligible compared to the local production (this is often the case in flow around obstacles), and it may be sufficient to impose inflow profiles constant in time. Similar remarks hold for DNS.

In regard to *initial* conditions, these partly depend on the spatio-temporal discretization method adopted for the solution of the flow and scalar transport equations. Some algorithms generate a sufficient amount of *numerical noise* to trigger turbulence during the early stages of a time-dependent simulation, even starting from smooth flow and scalar fields (e.g. zero velocities and uniform temperature), while, in other cases, it may be necessary explicit to introduce initial *perturbations*.

The prescription of perturbed initial conditions, especially in incompressible flow, depends on whether the numerical algorithms used do or do not withstand continuity violations (for example, most segregated algorithms of the SIMPLE family do). In the former case, it may be sufficient to superimpose on a smooth divergence-free initial distribution a pseudo-random perturbation field, without caring whether also this is divergence-free or not; continuity will be enforced anyway by the solver in the first time steps. In the latter case, it is necessary that the perturbation field, although pseudo-random, be divergence-free; a way of obtaining this is to generate a pseudo-random vector stream function and then compute the initial velocity perturbation as its rotor (which will necessarily be a divergence-free field).

References

Bardina J, Ferziger JH, Reynolds WC (1980) Improved subgrid-scale models for Large-Eddy simulation. AIAA Paper No. 80-1357

Breuer M, Rodi W (1994) Large-Eddy Simulation of turbulent flow through a straight square duct and a 180° bend. In: Voke PR, Kleiser L, Chollet J-P (eds) Direct and Large-Eddy simulation I. Kluwer, Dordrecht, pp 273–285

Ciofalo M (1994) Large-Eddy simulation: a critical survey of models and applications. In: Hartnett JP, Irvine TF Jr (eds) Advances in heat transfer, vol 25, chap 4,. Academic, New York, pp 321–419

Ciofalo M, Stasiek J, Collins MW (1996) Investigation of flow and heat transfer in corrugated passages—II. Numerical simulations. Int J Heat Mass Transfer 39(1):165–192

Davidson L (1997) Large Eddy simulation: a dynamic one–equation subgrid model for three–dimensional recirculation flow. In: Proceedings 11th symposium on turbulent shear flows. Grenoble, France, pp 26.1–26.6

Deardorff JW (1974) Three-dimensional numerical study of the height and mean structure of a heated planetary boundary layer. Boundary Layer Meteorol 7:81–106

Deardorff JW (1980) Stratocumulus-capped mixed layers derived from a three-dimensional model. Boundary Layer Meteorol 18:495–527

Fatica M, Orlandi P, Verzicco R (1994) Direct and large eddy simulations of round jets. In: Voke PR, Kleiser L, Chollet J-P (eds) Direct and Large-Eddy simulation I. Kluwer Academic Publishers, Dordrecht, pp 49–60

Garnier E, Adams N, Sagaut P (2009) Large Eddy simulations of compressible flows. Springer, Berlin

Gatski TB, Bonnet J-P (2013) Compressibility, turbulence and high speed flow, 2nd edn. Academic, New York

Gavrilakis S (1992) Numerical simulation of low-Reynolds number turbulent flow through a straight square duct. J Fluid Mech 244:101–129

Germano M (1992) Turbulence: the filtering approach. J Fluid Mech 238:325–336

Leonard A (1974) Energy cascade in Large Eddy simulation of turbulent fluid flows. Adv Geophys A 18:237–248

Lesieur M (1990) Turbulence in fluids. Kluwer Academic Publishers, Dordrecht

Lilly DK (1966) On the application of the eddy viscosity concept in the inertial subrange of turbulence. Report NCAR-123, National Center for Atmospheric Research, Boulder, Colo

Lilly DK (1992) A proposed modification of the Germano subgrid–scale closure method. Phys Fluids A 4:633–635

Mason PJ, Thomson DJ (1992) Stochastic backscatter in Large-Eddy simulations of boundary layers. J Fluid Mech 242:51–78

Piomelli U, Ferziger J, Moin P, Kim J (1989) New approximate boundary conditions for Large-Eddy simulations of wall-bounded flows. Phys Fluids A 1:1061–1068

Schmidt S, Thiele F (2002) Comparison of numerical methods applied to the flow over wall–mounted cubes. Int J Heat Fluid Flow 23:330–339

Schumann U (1975) Subgrid scale model for finite difference simulations of turbulent flows in plane channels and annuli. J Comp Phys 18:376–404

Smagorinsky J (1963) General circulation experiments with the primitive equations: Part I. Basic Experiment. Monthly Weather Rev 91:99–164

Spalart P, Jou WH, Strelets M, Allmaras S (1997), Comments on the feasibility of LES for wings, and on a hybrid RANS/LES approach. In: Liu C, Liu Z (eds) Advances in DNS/LES. Greyden Press, Columbus, OH

Speziale CG (1985) Galilean invariance of subgrid-scale stress models in the Large-Eddy Simulation of turbulence. J Fluid Mech 156:55–62

Van Driest ER (1956) On turbulent flow near a wall. J Aero Sci 23:1007–1011

Viswanathan AK, Tafti DK (2006) Detached Eddy simulation of turbulent flow and heat transfer in a two–pass internal cooling duct. Int J Heat Fluid Flow 27:1–20

Yoshizawa A (1982) A statistically-derived subgrid model for the Large-Eddy simulation of turbulence. Phys Fluids 25:1532–1538

Chapter 5
Rans Models

A numerical procedure without a turbulence model stands in the same relation to a complete calculation scheme as an ox does to a bull
Peter Bradshaw

Abstract Reynolds averaging is presented and the resulting Reynolds equations, containing unknown turbulent stresses and fluxes, are derived. Eddy viscosity models are classified and two of them, the k–ε and k–ω models, are presented in detail together with their variants such as low-Re k–ε models, RNG k–ε and SST. Exact Reynolds stress transport equations are also presented and a particular approximation of their unresolved terms, leading to a closed system of Reynolds stress transport equations, is illustrated. The application of RANS models to problems of fundamental and industrial interest is documented by several examples.

Keywords Turbulence modelling · Reynolds averaging · Eddy viscosity · K-epsilon · K-omega · Reynolds stress model

5.1 Reynolds Averaging

Unlike the LES equations discussed in the previous chapter, which are filtered in *space*, RANS (*Reynolds Averaged Navier–Stokes*) equations, by far the commonest approach to turbulence modelling, are usually regarded as *time*-filtered versions of the primitive continuity, momentum and scalar transport (in particular, energy) equations.

In principle, *finite-width* time filters $G(t, t')$ could be introduced, corresponding to the finite-width spatial filters used in LES. However, very little work has been done in this direction, see e.g. Collins et al. (1998), and in practice the only filter used is *long-term time averaging*, in which the generic turbulent quantity φ is decomposed into a *mean* component $\langle\varphi\rangle$ and a *fluctuating* component φ':

$$\langle\varphi(\mathbf{x}, t)\rangle = \lim_{t_{avg}\to\infty} \frac{1}{t_{avg}} \int\limits_{t}^{t+t_{avg}} \varphi(\mathbf{x}, t')dt' \tag{5.1}$$

© The Author(s), under exclusive license to Springer Nature Switzerland AG 2022
M. Ciofalo, *Thermofluid Dynamics of Turbulent Flows*, UNIPA Springer Series,
https://doi.org/10.1007/978-3-030-81078-8_5

$$\varphi'(\mathbf{x}, t) = \varphi(\mathbf{x}, t) - \langle\varphi(\mathbf{x}, t)\rangle \tag{5.2}$$

This approach is called *Reynolds decomposition*. Clearly, if the limit in Eq. (5.1) exists, it is independent from the initial instant t, so that the mean field loses its time-dependence and one can write $\langle\varphi(\mathbf{x}, t)\rangle = \langle\varphi(\mathbf{x})\rangle$, $\varphi'(\mathbf{x}, t) = \varphi(\mathbf{x}, t) - \langle\varphi(\mathbf{x})\rangle$.

The existence of the limit in Eq. (5.1) requires the flow to be *statistically stationary*, or *ergodic*. Therefore, *all* turbulence models stemming from Reynolds decomposition can rigorously be applied only to statistically stationary, albeit turbulent, flows (*steady turbulence*); their application to *unsteady turbulence*, in which the limit in Eq. (5.1) does not exist, is conceptually shaky and may result in severe errors.

Unsteady and transient turbulence will specifically be addressed in Chap. 7, where also alternative averaging approaches, including *phase-* and *ensemble-averaging*, will be introduced. Here, we will assume that the turbulent flow is, in a statistical sense, *stationary*, so that Eqs. (5.1), (5.2) can be applied. As in Chap. 4 (LES), we will also assume that *turbulent density fluctuations are negligible*.

Starting from the primitive continuity, Navier–Stokes and energy Eqs. (1.7), (1.17) and (1.28), proceed as done for LES in Chap. 4, i.e., substitute $\langle\varphi\rangle + \varphi'$ for the generic variable φ and then filter the resulting equations term by term. Time-averaged (Reynolds) equations are thus obtained which, apart from the different interpretation of the averaging operator $\langle\cdot\rangle$, are similar to the LES continuity, momentum and energy Eqs. (4.9), (4.10) and (4.17).

In Reynolds averaging, as in the case of LES filtering by finite-volume averaging, the definition of the averaging operator in Eq. (5.1) is such that Leonard and cross terms *vanish* and additional stresses and fluxes are simply given by:

$$\wp_{t,ij} = -\rho\langle u_i' u_j'\rangle \tag{5.3}$$

$$q_{t,i} = \rho c_p\langle u_i' T'\rangle \tag{5.4}$$

As in LES, the *closure* problem consists of expressing the terms $\wp_{t,ij}$ and $q_{t,i}$ as functions of average flow quantities.

5.2 Eddy Viscosity Models

These are based on a *gradient diffusion* assumption for additional stresses and fluxes (Boussinesq hypothesis), expressed by relations formally identical to Eqs. (4.18), (4.22) and (4.23):

$$\left[\wp_{t,ij} - \frac{1}{3}\delta_{ij}\wp_{t,kk}\right] = 2\mu_t\left[\langle S_{ij}\rangle - \frac{1}{3}\delta_{ij}\langle S_{kk}\rangle\right] \tag{5.5}$$

$$q_{t,i} = -c_p \Gamma_t \frac{\partial \langle T \rangle}{\partial x_i} \tag{5.6}$$

$$\Gamma_t = \frac{\mu_t}{\sigma_t} \tag{5.7}$$

The left hand side of Eq. (5.5) can also be written as $\tau_{t,ij}$ and represents the traceless part of the turbulent stress tensor $\wp_{t,ij}$. The resulting averaged momentum and energy equations (Reynolds equations) are:

$$\frac{\partial \rho \langle u_i \rangle}{\partial t} + \frac{\partial \rho \langle u_i \rangle \langle u_j \rangle}{\partial x_j} = -\frac{\partial p^{**}}{\partial x_i} + \frac{\partial}{\partial x_j}\left[(\mu + \mu_t)\left(\frac{\partial \langle u_i \rangle}{\partial x_j} + \frac{\partial \langle u_j \rangle}{\partial x_i} \right) \right] \tag{5.8}$$

$$\frac{\partial \rho \langle T \rangle}{\partial t} + \frac{\partial \rho \langle u_i \rangle \langle T \rangle}{\partial x_i} = \frac{\partial}{\partial x_i}\left[\left(\Gamma + \frac{\mu_t}{\sigma_t} \right) \frac{\partial \langle T \rangle}{\partial x_i} \right] \tag{5.9}$$

The quantities μ_t, Γ_t, σ_t are called *turbulent viscosity, turbulent heat diffusivity* and *turbulent Prandtl number*, respectively. As mentioned in Sect. 2.9, values between 0.8 and 0.9 are usually adopted for σ_t, independent of the molecular Prandtl number $\sigma = c_p \mu / \lambda$. The sum $\mu + \mu_t$ is called *effective*, or *total*, viscosity, and the sum $\Gamma + \mu_t/\sigma_t$ is called *effective*, or *total*, thermal diffusivity.

The effective pressure p^{**} includes the isotropic term $-1/3\wp_{t,kk}$ (implicit summation), opposite of the average of the diagonal terms in the turbulent stress tensor. This term can be regarded as a turbulent pressure p_t and can also be written as $2/3\rho k$ where $k = 1/2\langle u'_k u'_k \rangle$ is the *turbulent kinetic energy*. Therefore, one has $p^{**} = \langle p^* \rangle + 2/3\rho k$, in which p^* is the modified pressure defined as $p^* = p + 2/3\mu \nabla \mathbf{u}$ in Sect. 1.2.

In the end, p^{**} is a *second modified pressure* equal to $\langle p \rangle + \frac{2}{3}\mu \langle \nabla \cdot \mathbf{u} \rangle + \frac{2}{3}\rho k$. The difference between the effective pressure p^{**} and the thermodynamic pressure p can be a shortcoming only in the rare cases in which boundary conditions are known for p, but not for p^{**}, and is otherwise irrelevant: p^{**} is an unknown of the overall fluid dynamics problem exactly as p.

The heuristic arguments discussed in Sect. 2.9 suggest the turbulent viscosity μ_t to be expressed (apart from a numerical coefficient) as the product of density, eddy length scale and eddy velocity scale, Eq. (2.16). If Prandtl's mixing length concept is adopted, one has Eq. (2.21), in which the only quantity to be modelled is l. Elementary ways of doing this have been discussed at the end of Sect. 2.9.

Very simple turbulence models, based on the above or similar assumptions, are still used with some success in problems involving boundary or two-dimensional flows without recirculation, as often encountered in *turbomachinery* and *aeronautics*. For example, the popular model of Cebeci and Smith (1974) expresses μ_t as:

$$\mu_t = \rho(\kappa y f_\mu)^2 |\partial \langle u \rangle \partial y| \quad (y \leq y_c) \tag{5.10}$$

$$\mu_t = \rho K \gamma_I \delta_{BL} * u_e \quad (y > y_c) \tag{5.11}$$

Equation (5.10) holds in the near-wall region; here $\kappa \approx 0.42$ is the von Karman constant, y is the distance from the wall and f_μ is the Van Driest near-wall damping factor, Eq. (2.23). Equation (5.11) holds in the outer region; here $K \approx 0.0168$ is called the Clauser constant, $\gamma_I = [1 + 5.5(y/\delta_{BL})^6]^{-1}$ is the so called Klebanoff intermittence factor, δ_{BL} and $\delta_{BL}{}^*$ are kinematic thickness and momentum thickness of the boundary layer, and u_e is the principal velocity at the edge of the boundary layer. The value of y_c is that for which Eqs. (5.10) and (5.11) yield the same result.

Different extensions and re-formulations of the Cebeci-Smith model have been proposed, e.g. by Baldwin and Lomax (1978); see Chima et al. (1993) for applications to turbomachinery.

All the above models are *algebraic* models, which provide the turbulent viscosity as a function of mean flow quantities without using differential equations; when used in CFD codes, they cause a negligible increase of the computational load with respect to equivalent laminar problems. On the other hand, they are ad hoc models, optimized for specific problems only and containing a considerable number of "free" parameters which must be adjusted for each flow. Such models eventually rest on the theory of equilibrium boundary layers (Schlichting 1968), and are difficult to apply to three-dimensional problems with flow separation, re-attachment and recirculation.

A much greater generality is provided by *differential* models, in which the eddy viscosity is obtained from one or more quantities characteristic of turbulence (e.g. turbulent kinetic energy, dissipation or eddy length scale), which in their turn are the solutions of appropriate differential *transport equations*. Each transport equation expresses a *balance* between *generation, destruction, advective/diffusive transport* and (if appropriate, and within the limits discussed previously) *time variation*.

Apart from the model of Spalart and Allmaras (1992), in which the eddy viscosity is itself the solution of such a transport equation, most differential models include a transport equation for the turbulent kinetic energy $k = \frac{1}{2}\langle u_i' u_i' \rangle$. This has the form

$$\frac{\partial \rho k}{\partial t} + \frac{\partial \rho \langle u_i \rangle k}{\partial x_i} = \frac{\partial}{\partial x_j}\left[\left(\mu + \frac{\mu_t}{\sigma_k}\right)\frac{\partial k}{\partial x_j}\right] + P_k - \rho\varepsilon \qquad (5.12)$$

in which σ_k is a constant of unity order, called *turbulent Prandtl number for k*, while P_k and $\rho\varepsilon$ are the rates of production and dissipation of k per unit volume, both dimensionally equivalent to a volumetric power density ($\mathrm{Wm^{-3}}$, or $\mathrm{kg\ m^{-1}\ s^{-3}}$).

The production term, P_k, can be expressed in general flows as

$$P_k = -\wp_{t,ij}\frac{\partial \langle u_i \rangle}{\partial x_j} = \rho\langle u_i' u_j'\rangle\frac{\partial \langle u_i \rangle}{\partial x_j} \qquad (5.13)$$

i.e., as the contraction of the turbulent stress tensor $\wp_{t,ij}$ by the mean velocity gradients, changed in sign because of the definitions of k (positive) and of the normal Reynolds stresses (negative). P_k can be seen as the work done by the turbulent stresses on the mean flow per unit volume and per unit time (p.u.v.t.).

It can be demonstrated that, in constant-density fluids and under the eddy viscosity (Boussinesq, or gradient-diffusion) hypothesis, Eq. (5.5), the expression of P_k becomes

$$P_k = 2\mu_t \langle S_{ij} \rangle \langle S_{ij} \rangle \tag{5.14}$$

In the simultaneous presence of a body force (e.g. gravity) and of a density gradient (e.g. caused by a temperature gradient), a further contribution G_k to the production of turbulence arises, independent of mean shear. The influence of buoyancy production will be discussed in detail in Chap. 6. Here, we will assume that G_k is not present so that the production by shear, P_k, coincides with the total production of turbulence energy.

In regard to dissipation ε, this is formally defined, apart from the density factor ρ, as the second term in the right hand side of Eq. (2.3), i.e. $2\mu \langle S'_{ij} S'_{ij} \rangle$; since this expression contains fluctuating quantities S'_{ij}, it cannot be directly computed from mean (i.e., resolved) quantities, but a *closure model* is required.

According to Kolmogorov's theory of turbulence, a noteworthy relation exists between k, ε and the length scale l characteristic of turbulent eddies:

$$l = C_\mu \frac{k^{3/2}}{\varepsilon} \tag{5.15}$$

in which C_μ is a constant of the order of 0.09 (Landau and Lifschitz 1959).

Once both the turbulent kinetic energy k and the turbulence length scale l (or the dissipation ε) are known, the turbulent viscosity μ_t can be expressed, on the basis of the usual dimensional arguments, as $\mu_t \approx \rho l k^{1/2}$, or, taking account of Eq. (5.15):

$$\mu_t = \rho C_\mu \frac{k^2}{\varepsilon} \tag{5.16}$$

(*Prandtl-Kolmogorov* equation). Two main alternatives are now possible:

- **One-equation models**: in these models, the only transport equation is Eq. (5.12) for turbulent kinetic energy k, while the dissipation ε or the length scale l are *algebraically* prescribed as functions of the features of the mean flow. In the past, they found applications mainly in boundary layer problems, but were also extended, by means of more or less ad hoc prescriptions for l or ε, to more complex problems involving flow separation and recirculation (Thomas et al. 1981). The same reservations mentioned for purely algebraic models hold, albeit to a lesser extent, also for these models.
- **Two-equation models**: in these models, besides k, also a second quantity (dissipation ε, length scale l, eddy frequency $\omega = \varepsilon/k$, or others) is obtained by solving a differential transport equation which contains suitable production, destruction, advection/diffusion and time variation terms. This allows a greater generality and a lesser dependence from empirical assumptions, at the cost of a slightly larger computational load.

5.3   The k–ε Model and Its Variants

Among two-equation eddy diffusivity models, the one boasting by far the greatest success and the greatest longevity is the k–ε, first developed by the turbulence research group at the Imperial College in Londra (Launder and Spalding 1972) and then evolved into a large family of variants and generalizations. In the model's basic form, the second transport equation concerns the dissipation ε and takes the form

$$\frac{\partial \rho \varepsilon}{\partial t} + \frac{\partial \rho \langle u_i \rangle \varepsilon}{\partial x_i} = \frac{\partial}{\partial x_j}\left[\left(\mu + \frac{\mu_t}{\sigma_\varepsilon}\right)\frac{\partial \varepsilon}{\partial x_j}\right] + C_1 \frac{\varepsilon}{k}P - C_2 \rho \frac{\varepsilon^2}{k} \qquad (5.17)$$

in which σ_ε is a constant of unity order, called *turbulent Prandtl number for ε*.

Equation (5.17) is built after the corresponding k-Eq. (5.12), with ε production and destruction terms proportional to their k counterparts through the scaling factor ε/k and corrective coefficients C_1, C_2. A further constant (C_3) is required if dissipation by buoyancy has to be modelled, but this will omitted here for the sake of simplicity (see Chap. 6 for details). The *consensus* values for the various constants that appear in the model are $C_\mu = 0.09$, $C_1 = 1.44$, $C_2 = 1.92$, $\sigma_k = 1$, $\sigma_\varepsilon = 1.3$, obtained by a judicious mix of theoretical arguments, asymptotic considerations and comparisons of computational predictions against a vast body of experimental data or exact solutions.

Equations (5.12) and (5.17), together with the expression (5.13) or (5.14) for P_k, the Prandtl-Kolmogorov Eq. (5.16) for μ_t, and the above consensus values for the constants, define the "standard" k–ε model.

As an example of the results, Fig. 5.1 reports a mean velocity vector plot and the distribution of the three quantities k, ε and μ_t in a typical 3-D k–ε simulation; the configuration simulated represents a typical problem of industrial interest (transverse air flow across a bundle of tubes provided with plate fins; winglets oriented at 45° with respect to the main flow are also present to promote mixing). The plane visualized is midway between two plate fins.

The turbulent kinetic energy k takes its highest values in the high-shear regions because of the production term $P_k = 2\mu_t \langle S_{ij} \rangle \langle S_{ij} \rangle$. The dissipation ε is highest in the near-wall regions. The distribution of the resulting eddy viscosity μ_t follows approximately that of k, but with significant differences due to the concomitant effect of ε; in particular, as Eq. (5.16) shows, μ_t is small near walls, where the effective viscosity μ_{eff} practically reduces to the molecular viscosity μ.

In its fifty years of life, the k–ε model has found uncountable applications and is still the first option in most CFD codes. Among RANS models, it has been regarded for many years as the best compromise between generality, accuracy, ease of implementation and computational stability (Mohammadi and Pironneau 1994).

Among the most significant shortcomings of the k–ε model, some are intrinsic to its being an eddy viscosity model: these include, for example, the inability to predict secondary motions due to the anisotropy of normal turbulent stresses (Fig. 4.3).

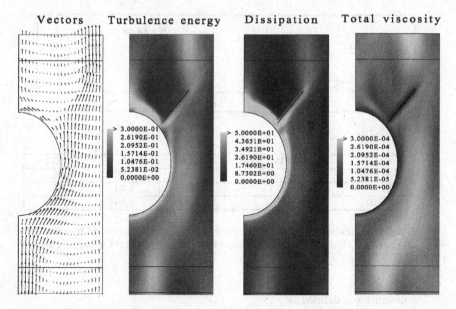

| Vectors | Turbulence energy | Dissipation | Total viscosity |

Fig. 5.1   3-D *k–ε* simulation of turbulent air flow across a bundle of tubes provided with orthogonal plate fins and mixing-promoting winglets at 45° with the main flow (author's own results)

Others arise from the difficulty to cover with a single set of constants the range of all possible turbulent flows; these include the systematic underestimation of the reattachment length in separated flow, of which the *backward-facing step* is the most illustrious example, and the overestimation of the rate of spreading of submerged jets. These problems are related to an unsatisfactory modelling of the ε production term in Eq. (5.17), which is assumed to be simply proportional to the corresponding k production term in Eq. (5.12). A suitable change of the constants can improve the aspects mentioned above, but would also impair the predictions in other regions of the flow and the model's ability to predict simpler flows.

An approach of greater generality consists of making the model "constants" functions of some feature of the mean flow. A large number of modifications have been proposed to account for streamline curvature, recirculation, irrotational strains, turbulence anisotropy and other features that differentiate complex flows from simple boundary layer flows (Hanjalić and Launder 1980; Chen and Kim 1987).

Among the proposed modifications, one that rests on a sound theoretical basis is the so called RNG (*Re-Normalization Group*) *k–ε* model (Yakhot et al. 1992). The model has its roots in the re-normalization group theory (Yakhot and Orszag 1986) but ultimately results in expressing the coefficient C_2 in Eq. (5.17), which is a constant equal to 1.92 in the standard *k–ε* model, as

$$C_2 = C_2^0 + \frac{C_\mu \dot{\eta}^3 (1 - \eta/\eta_0)}{1 + \beta_{RNG}\eta^3} \qquad (5.18)$$

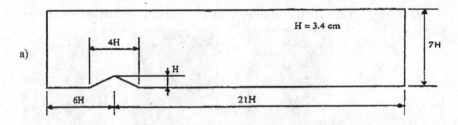

a)

b)

CASE	EXPERIMENTAL	k-ε	DS	RNG k-ε
NO-BL	10-11H	6.85H	7.1H	8.8H
BL	7-8H	6.2H	6.4H	7.4H
GRID	7-8H	6.7H	6.85H	7.8H

Fig. 5.2 Turbulent flow with separation over a triangular obstacle. **a** geometry; **b** reattachment lengths as measured or predicted by three turbulence models. From results of Ciofalo and Palagonia (1996)

The quantity η is defined as

$$\eta = \frac{Sk}{\varepsilon} = \frac{S}{\omega} \tag{5.19}$$

in which $S = \left(2\langle S_{ij}\rangle\langle S_{ij}\rangle\right)^{1/2}$. Thus, η is the ratio of the irrotational to the rotational component of the velocity gradient. β_{RNG} and η_0 are two new constants for which Yakhot et al. suggest the values of 0.012 and 4.38, respectively. The values recommended for the other constants are $C_\mu = 0.0845$, $\sigma_k = \sigma_\varepsilon = 0.7194$, $C_1 = 1.42$, $C_2{}^0 = 1.68$, all somewhat different from the values adopted in the standard k–ε model.

An example of the improvements allowed by the RNG k–ε model in separated flows is provided in Fig. 5.2 (Ciofalo and Palagonia 1996). Table (b) compares values of the reattachment length downstream of a triangular obstacle, sketched in graph (a), for three different inlet conditions dubbed "NO-BL", "BL" and "GRID".

The values reported are expressed as multiples of the obstacle height H. Experimental data were obtained by Laser-Doppler anemometry in a wind tunnel at the Von Karman Institute in Bruxelles. Predictions were obtained by the CFDS-FLOW3D code (Burns et al. 1989), an ancestor of the current ANSYS-CFX® code, using three different turbulence models: standard k–ε, RNG k–ε and a more complex differential stress model (DS) of the kind that will be discussed in Sect. 5.8.

The RNG k–ε model yields reattachment lengths in better agreement with experimental data with respect not only to the standard k–ε model but also to the second-order differential stress model. A further advantage of the RNG k–ε model is that, unlike the standard model, it can be extended to the quasi-laminar near-wall regions, because the parametrization of the ε production term, expressed by Eqs. (5.16), (5.17), implicitly accounts for the enhanced dissipation observed in these regions.

Therefore, the model does not need the *wall functions* described in the following Sect. 5.4 and thus behaves, under this respect, like one of the low Reynolds number models discussed in Sect. 5.5 below.

5.4 Boundary Conditions and Wall Functions

Neither the transport equations for k and ε, Eqs. (5.12) and (5.17), nor the Prandtl-Kolmogorov Eq. (5.16), expressing the turbulent viscosity μ_t as a function of these two quantities, are strictly applicable in the near-wall region, in particular in the viscous/conductive sublayer. In the context of the $k–\varepsilon$ family models, this problem has received two alternative solutions:

- *wall functions*: the computational grid does not resolve the viscous/conductive sublayer, but the expressions that link velocity and temperature at the grid point closest to the wall with shear stress and heat flux at the wall are reformulated so as to account for the interposed presence of the sublayer;
- so called *low Reynolds number models*: the computational grid does resolve the viscous/conductive sublayer, but the transport equations for k and ε and the Prandtl-Kolmogorov equation are reformulated so as to hold also in this region.

The former approach will be discussed in this Section, the latter in the following. In all turbulent flows, one can assume that a viscous/conductive sublayer exists near all solid walls, such that in this region momentum, heat and mass transfer are controlled by the respective molecular diffusivities and all velocity, temperature and concentration profiles are linear with the distance y from the wall (Arpaci and Larsen 1984; Hinze 1975). This sublayer is followed by an outer region of the boundary layer, characterized by profiles that vary logarithmically with y; an intermediate buffer layer is sometimes assumed to exist between inner and outer regions.

Using the friction velocity $u_\tau = (\tau_w/\rho)^{1/2}$ as the velocity scale, ν/u_τ as the length scale and—if appropriate—$q_w/(\rho c_p u_\tau)$ as the temperature scale (*wall scales*), the dimensionless variables $y^+ = y u_\tau/\nu$, $u^+ = u/u_\tau$, $T^+ = (T - T_w^-)\rho c_p u_\tau/q_w$ can be defined.

For *smooth walls*, the wall-parallel velocity component follows the universal profile:

$$u^+ = y^+ \quad (y^+ \leq y_v^+) \tag{5.20}$$

$$u^+ = \ln(Ey^+)/\kappa \quad (y^+ > y_v^+) \tag{5.21}$$

in which $\kappa \approx 0.42$ is the *von Karman* constant, $y_v^+ \approx 11.2$ is the dimensionless thickness of the viscous sublayer and $E \approx 9.86$, determined by imposing that Eqs. (5.20) and (5.21) yield the same value of u^+ for $y^+ = y_v^+$.

Temperature follows the universal profile:

$$T^+ = \sigma y^+ \quad (y^+ \le y_T^+) \tag{5.22}$$

$$T^+ = \sigma_t[\ln(Ey^+)/\kappa + P_J] \quad (y+ > y_T^+) \tag{5.23}$$

in which σ is the Prandtl number $c_p\mu/\lambda$, σ_t is the turbulent Prandtl number introduced in Eq. (5.9), usually assumed equal to 0.8–0.9, and P_J is a function of both Prandtl numbers. This dependence was experimentally investigated in detail by Jayatilleke (1969) in a broad range of Prandtl numbers from 0.3 to 10^4, and for smooth walls can be expressed as

$$P_J = 9.24\left[\left(\frac{\sigma}{\sigma_t}\right)^{3/4} - 1\right]\left[1 + 0.28 \cdot \exp\left(-0.007\frac{\sigma}{\sigma_t}\right)\right] \tag{5.24}$$

In Eqs. (5.22), (5.23), $y_T{}^+$ represents the dimensionless thickness of the conductive sublayer and is determined by imposing that, for $y^+ = y_T{}^+$, the two expressions yield the same value of T^+.

Equations (5.22), (5.23) are written for temperature T, but can be extended in an obvious way to a generic scalar φ. In this case the Prandtl number σ will have to be replaced by the Schmidt number Sc.

The resulting universal velocity and temperature profiles in the wall region are reported in Fig. 5.3 below. The curve for $\sigma = 1$ practically coincides with the universal

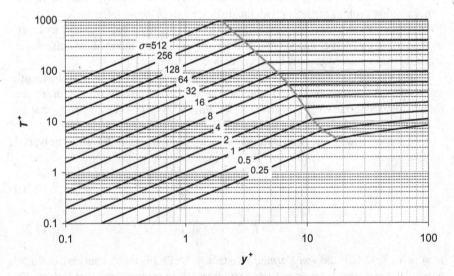

Fig. 5.3 Universal near-wall temperature profiles $T^+(y^+)$ for different values of the Prandtl number σ according to the experimental results of Jayatilleke (1969). The curve for $\sigma = 1$ represents also the universal velocity profile $u^+(y^+)$. The thick grey line is the locus of the condition $y^+ = y_T{}^+$

velocity profile in Eqs. (5.20), (5.21).

The thick grey line in Fig. 5.3 is the locus of the points where $y^+ = y_T^+$ for any given σ. It can be observed that the dimensionless thickness y_T^+ of the conductive sublayer decreases as the Prandtl number increases; for example, it is 11.2 (equal to y_v^+) for $\sigma = 1$, but only ~ 2 for $\sigma = 500$. This behaviour is consistent with the remarks on the minimum spatial scale of inhomogeneities in a scalar distribution (Batchelor length) made in Sect. 2.5. However, the Batchelor scale decreases as $\sigma^{-1/2}$, while y_T^+ decreases less steeply with σ as Fig. 5.3 shows.

Now, let u_P be the velocity component parallel to a wall at a grid point P located at a distance y_P from the wall. Equations (5.20), (5.21) can be regarded as equations in u_τ:

$$u_\tau = (v u_P / y_P)^{1/2} \quad \text{if } u_P y_P \leq v(y_v^+)^2 \tag{5.25}$$

$$u_\tau = u_P / [\ln(E y_P u_\tau / v) / \kappa] \quad \text{if } u_P y_P > v(y_v^+)^2 \tag{5.26}$$

from which the friction velocity u_τ and the wall shear stress $\tau_w = \rho u_\tau^2$ can be evaluated. If the first grid point lies outside of the viscous sublayer, then Eq. (5.26) holds. This is a trascendent equation in u_τ which can be solved iteratively. The resulting relation between τ_w and u_P replaces the viscous formula (5.25), which can be written $\tau_w = \mu u_P / y_P$ and holds for laminar flow or for near-wall points P lying within the viscous sublayer. Similar considerations and formulae hold for the relation between near-wall temperature T_P and wall heat flux q_w.

In a numerical simulation, *both* u_P and τ_w are unknowns of the overall thermofluid dynamics problem. The nonlinear relation between τ_w and u_P expressed by Eq. (5.26) does not lend itself well to numerical solution algorithms and would require, for example, *deferred correction* techniques, in which the wall shear stress at the generic iteration would be computed from the velocity at the previous iteration.

In k–ε family turbulence models, the above difficulties can be overcome by adopting as the velocity scale in universal profiles not the friction velocity u_τ, but the root mean square velocity fluctuation at point P, expressed as proportional to $k_P^{1/2}$ (Launder and Spalding 1974). The rationale for this choice, which apparently makes a "universal" profile dependent on the position of a grid point, is that in equilibrium turbulent boundary layers a fairly broad region exists (between $y^+ \approx 20$ and $y^+ \approx 60$), in which k is close to $\tau_w/(\rho C_\mu^{1/2}) \approx 3.33 u_\tau^2$. Therefore, provided the grid point P lies within this region, $u_\tau = (\tau_w/\rho)^{1/2}$ and $u^* = C_\mu^{1/4} k_P^{1/2} \approx 0.548 k_P^{1/2}$ are equivalent scales.

If this approach is adopted, Eqs. (5.20), (5.21) remain formally valid, but the dimensionless quantities y^+, u^+ are redefined as $y^+ = y u^*/v$, $u^+ = u u^*/(\tau_w/\rho)$, $T^+ = (T - T_w)\rho c_p u^*/q_w$. The relation between velocity in P and wall shear stress becomes

$$\tau_w = (\mu / y_P) u_P \quad \text{if } y_P u * / v \leq y_v^+ \tag{5.27}$$

$$\tau_w = \rho u * u_P/[\ln(E y_P u * /v)/\kappa] \quad \text{if } y_P u * /v > y_v^+ \tag{5.28}$$

and thus is *linear* with respect to u_P even if P lies in the logarithmic (outer) region of the wall layer. This greatly simplifies the solution of the discretized flow equations.

By similar arguments one can derive the new relation linking the temperature at point P with the wall heat flux; the treatment can immediately be extended to other passive scalars. This scaling method is adopted in most CFD codes when the k–ε or similar turbulence models are used.

The use of wall functions and universal profiles in problems more complex than equilibrium boundary layers (e.g. problems involving section changes, adverse pressure gradients, separation and reattachment) is questionable. The most critical issue is the modelling of the reattachment region of separated shear layers, where the wall shear stress and the near-wall parallel velocity vanish altogether, while velocity fluctuations attain their highest values: in these regions, u_τ or u^* are obviously inadequate scales for velocity profiles. Several modifications and adaptations of the wall function method to these and other critical conditions have been proposed in the literature, see e.g. Chieng and Launder (1980) or Cruz and Silva-Freire (1998).

As an example, Fig. 5.4 (Ciofalo and Collins 1989) reports streamwise profiles of the local Nusselt number (Nu) along a wall following a backward-facing step, a classic problem in turbulence modelling. The abscissa is the distance from the step, normalized by the reattachment length to exclude the effect of the systematic underestimation of this quantity by k–ε family models. The reported *Nu* profiles include experimental data by Vogel and Eaton (1985) (symbols), predictions provided by the k–ε model using standard wall functions (line a) and predictions based on

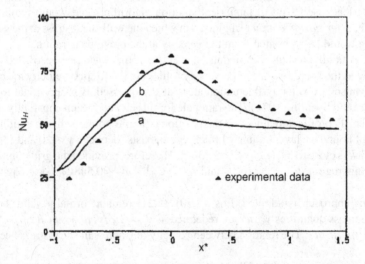

Fig. 5.4 Local Nusselt number on the wall that follows a backward-facing step. Symbols: experimental data (Vogel and Eaton 1985); **a:** k–ε model with standard wall functions; **b:** k–ε model with modified wall functions. From results of Ciofalo and Collins (1989)

modified wall functions proposed by the authors (line b). The Reynolds number (Re), based, like Nu, on the step height, was 28,000. The modifications consisted of making the nondimensional thickness of the viscous sublayer, y_v^+, vary as a function of the local turbulence intensity, k_P/u_P^2. Standard wall functions yield a severe underestimation of the experimentally observed Nu peak at the reattachment point ($x^* = 0$); predictions improve considerably when the modified wall functions are used.

More complex wall functions have also been proposed, based on modelling the near-wall region as composed of more than just two layers; for example, Amano et al. (1983) distinguish a viscous sublayer for $y^+ \leq 5$, a fully turbulent outer region for $y^+ \geq 30$, and an intermediate *buffer layer* for $5 < y^+ \leq 30$.

5.5 Low Reynolds Number k–ε Models

With respect to using wall functions, a more radical method of dealing with the wall regions of turbulent flows are the so-called *low Reynolds number turbulence models*. Several models of this family can be regarded as modified versions of the k–ε model. In all of them, the Prandtl-Kolmogorov Eq. (5.16) is re-written as

$$\mu_t = \rho f_\mu C_\mu k^2 / \varepsilon \tag{5.29}$$

which contains a damping factor f_μ accounting for the reduction of the spatial turbulence scales caused by the proximity of solid walls. In addition, the transport equation for dissipation is often reformulated in terms of the auxiliary variable $\varepsilon^* = \varepsilon - D_\varepsilon$, so that Eq. (5.17) becomes

$$\frac{\partial \rho \varepsilon^*}{\partial t} + \frac{\partial \rho \langle u_i \rangle \varepsilon^*}{\partial x_i} = \frac{\partial}{\partial x_i} \left[\left(\mu + \frac{\mu_t}{\sigma_\varepsilon} \right) \frac{\partial \varepsilon^*}{\partial x_i} \right] + f_1 C_1 \frac{\varepsilon^*}{k} P - f_2 C_2 \rho \frac{\varepsilon^{*2}}{k} + \rho E_\varepsilon \tag{5.30}$$

Different proposed low Reynolds number turbulence models mainly differ in the form taken by the functions $f_1, f_2, f_\mu, D_\varepsilon$ and E_ε. For example, such expressions are summarized in Table 5.1 for the models proposed by Launder and Sharma (1974), Lam and Bremhorst (1981) and Nagano and Hishida (1987).

In all cases, the two auxiliary Reynolds numbers R_t, R_k in the expressions for f_2 and f_μ are defined as

$$R_t = k^2 / (v\varepsilon) \tag{5.31}$$

$$R_k = k^{1/2} y / v \tag{5.32}$$

Table 5.1 Constants and functions in three low-Reynolds number k–ε models

	Lam and Bremhorst	Launder and Sharma	Nagano and Hishida
f_1	$1 + \left(\frac{0.005}{f_\mu}\right)^3$	1	1
f_2	$1 - \exp(-R_t^2)$	$1 - 0.3\exp(-R_t^2)$	$1 - 0.3\exp(-R_t^2)$
f_μ	$[1 - \exp(-0.0165R_k)]^2$ $\times \left(1 + \frac{20.5}{R_t}\right)$	$\exp\left[-\frac{3.4}{\left(1+\frac{R_t}{50}\right)^2}\right]$	$\left[1 - \exp\left(-\frac{y^+}{26.5}\right)\right]^2$
D_ε	0	$2\nu\left(\frac{\partial k^{1/2}}{\partial x_j}\right)\left(\frac{\partial k^{1/2}}{\partial x_j}\right)$	$2\nu\left(\frac{\partial k^{1/2}}{\partial x_j}\right)\left(\frac{\partial k^{1/2}}{\partial x_j}\right)$
E_ε	0	$2\nu\nu_t\left(\frac{\partial^2 u_i}{\partial x_j \partial x_k}\right)\left(\frac{\partial^2 u_i}{\partial x_j \partial x_k}\right)$	$2\nu\nu_t(1 - f_\mu)\left(\frac{\partial^2 u_i}{\partial x_j \partial x_k}\right)\left(\frac{\partial^2 u_i}{\partial x_j \partial x_k}\right)$

in which y is the distance from the nearest solid wall. The dimensionless coordinate y^+ is defined as $y^+ = yu_\tau/\nu$ or as $y^+ = yC_\mu^{1/4}k_P^{1/2}/\nu$.

In all low Reynolds number turbulence models, the viscous and conductive sublayer must be resolved by the computational grid, and it is advisable that it includes several grid points (e.g. 10–15).

The no slip conditions, yielding $\tau_w = \mu u_P/y_P$, are directly imposed at the first near-wall grid point (no wall functions). The wall boundary condition for the turbulent kinetic energy k is usually the simple Dirichlet condition $k_w = 0$.

In regard to dissipation, in models using the standard definition of ε, i.e. assuming $D = 0$, as the Lam and Bremhorst model, the wall boundary condition for this variable may be a bit problematic; a common choice is the Neumann condition $\nabla_\varepsilon \cdot \mathbf{n}|_w = 0$ (i.e., zero normal derivative of ε at solid walls). In models adopting the modified variable $\varepsilon^* = \varepsilon - D$, the Dirichlet condition $\varepsilon^*|_w = 0$ is usually imposed.

Several comparative studies of low Reynolds number turbulence models have appeared in the literature. For example, Patel et al. (1985) evaluated eight such models (including those in Table 5.1) for their ability to reproduce experimental results concerning different turbulent boundary layer problems (flat plate, adverse pressure gradient, favourable pressure gradient with re-laminarization).

Moving from simple boundary layer problems to more complex flows (exhibiting, for example, recirculation, natural convection and three-dimensionality), the performance of low-Re models may change, and such issues as robustness, ease of implementation and computing effort may become crucial.

In the author's experience, overall satisfactory results were obtained by using the models of Launder and Sharma (1974) and Lam and Bremhorst (1981).

For example, Fig. 5.5 reports results for heat transfer between a forced air flow and a horizontal flat wall in the presence of a second, vertical, wall bearing transverse ribs as turbulence promoters. Predictions obtained by the Launder and Sharma model, as implemented in the computer code CFX-4 (AEA Technology 1994), are compared with experimental results of Tanda et al. (1995) using liquid crystals thermography.

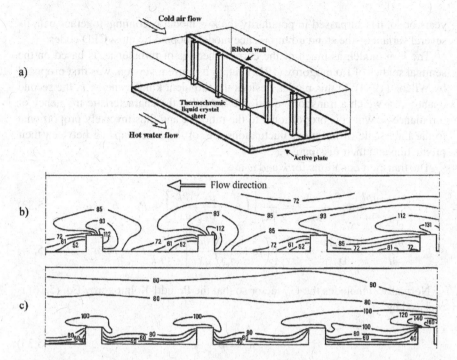

Fig. 5.5 Heat transfer from a flat wall in the presence of an orthogonal ribbed wall. a) schematic of the test section; b) experimental isolines of the heat transfer coefficient in $W/(m^2 K)$; c) corresponding predictions obtained by the low-Re $k–\varepsilon$ model of Launder and Sharma (1974). From results of Tanda et al. (1995)

The agreement is satisfactory in regard to the overall levels of the heat transfer coefficient, less so in regard to the detailed shape of its isolines (which obviously reflect the flow structure in the proximity of the flat wall). Nevertheless, the low-Re model provided better results than either the standard $k–\varepsilon$ model with wall functions or a differential Reynolds stress–Reynolds flux model (see Sect. 5.8).

5.6 The $k–\omega$ Model and Its Variants

The two-equation $k–\varepsilon$ model is based on the choice of the turbulent kinetic energy per unit mass (k) and of its dissipation rate (ε) as the quantities characterizing turbulence. The quantity k characterizes the intensity of the turbulent fluctuations while ε characterizes their time scale (proportional to k/ε) or spatial scale (proportional to $k^{3/2}/\varepsilon$). Several two-equation models have been proposed in which the dissipation ε is replaced by some other variable as the quantity characterizing the turbulence scales. In the following, we will describe the $k–\omega$ model, which, in the last twenty years or so, has surpassed in popularity the k–ε model becoming–together with its several variants—the standard turbulence model adopted by most CFD codes.

The k–ω model, as much of the current theories of turbulence, is based on the seminal studies of Kolmogorov (1942) but, in its current version, was first proposed by Wilcox (1988). In this model, besides the turbulent kinetic energy k, the second quantity for which a transport equation is solved is the characteristic frequency of turbulence ω, which is proportional to the ratio ε/k and thus inversely proportional to the time scale of turbulent fluctuations, i.e. of the time elapsing between their production and their dissipation.

The transport equations for k and ω are

$$\frac{\partial \rho k}{\partial t} + \frac{\partial \rho \langle u_i \rangle k}{\partial x_i} = \frac{\partial}{\partial x_j}\left[\left(\mu + \frac{\mu_t}{\sigma_k}\right)\frac{\partial k}{\partial x_j}\right] + P_k - \beta^* \rho k \omega \tag{5.33}$$

$$\frac{\partial \rho \omega}{\partial t} + \frac{\partial \rho \langle u_i \rangle \omega}{\partial x_i} = \frac{\partial}{\partial x_j}\left[\left(\mu + \frac{\mu_t}{\sigma_\omega}\right)\frac{\partial \omega}{\partial x_j}\right] + \frac{5}{9}\frac{\omega}{k}P_k - \beta \rho \omega^2 \tag{5.34}$$

Normally ω includes the C_μ factor so that the Prandtl-Kolmogorov Eq. (5.16) is replaced by

$$\mu_t = \rho \frac{k}{\omega} \tag{5.35}$$

The production by shear, P_k, is as in the k–ε model, so that it can be written as $2\mu_t \langle S_{ij} \rangle \langle S_{ij} \rangle$. For the various constants, Wilcox recommends $\beta = 0.075$, $\beta^* = C_\mu = 0.09$, $\sigma_k = \sigma_\omega = 2$.

With respect to k–ε, the k–ω model offers the advantage of remaining valid also in the near-wall region, so that this can be resolved by the computational grid and treated as any other region of the computational domain, as in low Reynolds number turbulence models. On the other hand, the k–ω model suffers from its own shortcomings, such as an excessive sensitivity to turbulence intensity in the free-stream region and some overestimation of k in stagnation regions.

In order to overcome the limitations of the k–ω model, Menter (1993, 1994) proposed a combination of k–ω model in the near-wall region and and k–ε model (reformulated in terms of k and ω instead of k and ε) in the free stream. The two models are weighted by F_1 and $1 - F_1$, respectively, F_1 being a *blending function* which ranges from 1 at walls (yielding the k–ω model) to 0 in the free stream (yielding the k–ε model). The resulting model was dubbed SST (*Shear Stress Transport*).

Because of its widespread adoption, we will describe the SST model in detail, in the original 1993–1994 formulation. The transport equations for k and ω are written

$$\frac{\partial \rho k}{\partial t} + \frac{\partial \rho \langle u_j \rangle k}{\partial x_j} = \frac{\partial}{\partial x_j}\left[(\mu + c_k \mu_t)\frac{\partial k}{\partial x_j}\right] + P_k - \beta^* \rho k \omega \tag{5.36}$$

$$\frac{\partial \rho \omega}{\partial t} + \frac{\partial \rho \langle u_i \rangle \omega}{\partial x_i} = \frac{\partial}{\partial x_j}\left[(\mu + c_\omega \mu_t)\frac{\partial \omega}{\partial x_j}\right] + \frac{\rho \gamma}{\mu_t}P_k - \beta \rho \omega^2$$

$$+ 2(1 - F_1)\rho \frac{c_{\omega 2}}{\omega}\frac{\partial k}{\partial x_j}\frac{\partial \omega}{\partial x_j} \tag{5.37}$$

while the Prandtl-Kolmogorov Eq. (5.14) or (5.33) is re-formulated as

$$\mu_t = \frac{\rho a_1 k}{\max(a_1 \omega, \Omega F_2)} \tag{5.38}$$

in which

$$\Omega = \left(2\langle \Omega_{ij}\rangle\langle \Omega_{ij}\rangle\right)^{1/2} \tag{5.39}$$

(quadratic invariant of the average vorticity tensor, see Appendix).

The *blending function* F_1 is defined as

$$F_1 = \tanh\left(\left\{\min\left[\max\left(\frac{\sqrt{k}}{\beta^* \omega y}, \frac{500\mu}{\rho \omega y^2}\right), \frac{4\rho c_{\omega 2}k}{CD_{k\omega}y^2}\right]\right\}^4\right) \tag{5.40}$$

and vanishes far from solid walls (k–ε model), while tending to 1 in the near-wall region (k–ω model). The functions F_2 and $CD_{k\omega}$ are defined as

$$F_2 = \tanh\left\{\left[\max\left(\frac{2\sqrt{k}}{\beta^* \omega y}, \frac{500\mu}{\rho \omega y^2}\right)\right]^2\right\} \tag{5.41}$$

$$CD_{k\omega} = \max\left(2\rho \frac{c_{\omega 2}}{\omega}\frac{\partial k}{\partial x_i}\frac{\partial \omega}{\partial x_i}, 10^{-20}\right) \tag{5.42}$$

Note that c_k and c_ω are the *reciprocals* of turbulent Prandtl numbers for k and ω. The production by shear, P_k, is (at least in this version of the SST model) the same as in the k–ε and k–ω models, and can be written as $P_k = 2\mu_t \langle S_{ij}\rangle\langle S_{ij}\rangle$. Finally, $a_1 = 0.31$ and $\beta^* = 0.09$ while the remaining constants β, c_k and c_ω are linear combinations of those used for k–ω (subscript 1) and k–ε (subscript 2):

- $\beta_1 = 0.075$, $c_{k1} = 0.85$, $c_{\omega 1} = 0.65$;
- $\beta_2 = 0.0828$, $c_{k2} = 1$, $c_{\omega 2} = 0.856$.

weighted by the blending functions F_1 and $1 - F_1$, respectively. A slightly modified version of the model (Menter et al. 2003) was introduced by the author ten years after the first proposal; it will not be discussed here for the sake of simplicity.

As an example from the author's own experience with the SST model (Di Piazza and Ciofalo 2010), Fig. 5.6 reports the Darcy friction coefficient f_D in helical pipes.

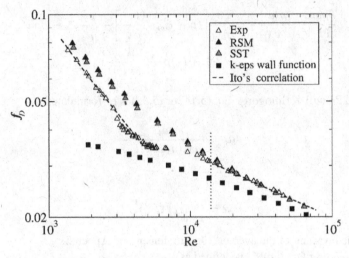

Fig. 5.6 Experimental and computational results for the Darcy friction coefficient f_D in a helical pipe with curvature 0.01. From results of Di Piazza and Ciofalo (2010)

The values of f_D computed by three different RANS models (the k–ε model with wall functions, the SST model and a differential Reynolds stress model, or RSM) as functions of the Reynolds number are compared with experimental results and with a well established empirical correlation in a broad range of Reynolds numbers, from steady laminar flow (Re < 3000) to fully turbulent flow (Re > 10,000).

SST predictions are close to those of the more complex RSM and, for Re > 10,000, also to the experimental results; they are acceptable also in the laminar region, showing that SST can predict laminarization, and less satisfactory only in the (difficult) transitional region 3000 < Re < 10,000. Contrariwise, the k–ε model with wall functions yields a wrong trend in the laminar region and underestimates f_D rather heavily also in the fully turbulent region.

5.7 Algebraic Reynolds Stress/Flux Models

A limitation shared by all turbulent viscosity models is that, as Eq. (5.5) shows, the deviatoric parts of the turbulent stress and mean strain rate tensors are proportional and thus necessarily aligned. Moreover, the three normal turbulent stresses $\langle u'^2 \rangle$, $\langle v'^2 \rangle$, $\langle w'^2 \rangle$ are not individually predicted, but rather included in a single scalar (the turbulent kinetic energy k); therefore, such phenomena as the secondary flows in non-circular ducts, which arise exactly from the asymmetry between these stresses, cannot be predicted.

The only possible remedy to these shortcomings is to abandon the concept of a turbulent viscosity (i.e., the so called *gradient diffusion*, Boussinesq-like, hypothesis) and to model the six independent turbulent stresses individually. This approach gives rise to the Reynolds stress models of turbulence.

In a first class of such models, the turbulent stresses $\wp_{t,ij}$ are modelled as prescribed *algebraic* functions of the mean flow field (and, in some versions, also of k and ε or ω, for which transport equations similar to Eqs. (5.12)–(5.15) or (5.34), (5.35) are retained). Models in this class are called *Algebraic Stress Models* (ASM).

Among the many Algebraic Stress Models proposed in the literature, we will discuss here the so called nonlinear k–ε model (Speziale 1987), which stands out for the depth and soundness of its theoretical basis. The author starts from the assumption that the traceless part of the stress tensor, $\tau_{t,ij} = \wp_{t,ij} - (1/3)\delta_{ij}\,\wp_{t,kk}$, can be expressed as a function of the mean velocity gradients, their *total* derivative and the scalars k, ε and ρ (of course, this last quantity is unnecessary in constant-density fluids):

$$\tau_{t,ij} = f\left(\frac{\partial \langle u_k \rangle}{\partial x_l};\ \frac{D}{Dt}\frac{\partial \langle u_k \rangle}{\partial x_l};\ k;\ \varepsilon;\ \rho\right) \tag{5.43}$$

(with the indexes k and l running, in general, over all possible values from 1 to 3).

The k–ε model can be regarded as the Taylor expansion of this expression, truncated after the first term. Developing the series to the second term, under very general assumptions (such as the positivity of k, the Galilean invariance of $\tau_{t,ij}$ and the correct asymptotic behaviour of $\tau_{t,ij}$ in flows subjected to rapid rotation) leads to the following expression:

$$\tau_{t,ij} = C\rho\frac{k^2}{\varepsilon}\langle S_{ij} \rangle + C^2\rho\frac{k^3}{\varepsilon^2}$$
$$\times \left[C_D\left(\langle S_{im}\rangle\langle S_{mj}\rangle - \frac{1}{3}\delta_{ij}\langle S\rangle^2\right) + C_E\left(\langle D_{ij}\rangle - \frac{1}{3}\delta_{ij}\langle D_{kk}\rangle\right) \right] \tag{5.44}$$

in which $\langle S_{ij} \rangle$ is the mean strain rate tensor, $\langle S \rangle^2$ is its quadratic invariant and $\langle D_{ij} \rangle$ is its *Oldroyd derivative*, defined as

$$\langle D_{ij} \rangle = \left[\frac{\partial}{\partial t}\langle S_{ij} \rangle + \langle u_k \rangle\frac{\partial}{\partial x_k}\langle S_{ij} \rangle\right] - \left[\frac{\partial \langle u_i \rangle}{\partial x_k}\langle S_{kj} \rangle + \frac{\partial \langle u_j \rangle}{\partial x_k}\langle S_{ki} \rangle\right] \tag{5.45}$$

(a tensor operator commonly used in the rheology of non-Newtonian fluids, but rarely in studies on the fluid dynamics of Newtonian fluids and turbulence).

For constant-density fluids, the first term at the RHS of Eq. (5.44) coincides with the Prandtl-Kolmogorov Eq. (5.16) used in eddy-viscosity models, provided $C = 2C_\mu$. The remaining terms are the nonlinear contributions that characterize the model. For both constants C_D, C_E Speziale suggests the value 1.68. In the cited paper, the author demonstrates that the nonlinear k–ε model allows the correct prediction of the secondary flows in the turbulent, fully developed motion of a fluid through rectangular ducts, as well as of the reattachment length in problems involving recirculation such as the classic backward-facing step; for both these problems, the standard, or *linear*, k–ε model errs.

In regard to the transport of enthalpy and other scalars, in the context of algebraic stress models the Gradient Diffusion Hypothesis (GDH), expressed by Eq. (5.6), can consistently be replaced by the so called *Generalized Gradient Diffusion Hypothesis* (GGDH), which can be expressed as

$$q_i = -C_\theta \frac{k}{\varepsilon} \rho c_p \langle u_i' u_j' \rangle \frac{\partial \langle T \rangle}{\partial x_j} \tag{5.46}$$

in which, with respect to Eq. (5.6), the *scalar* diffusivity Γ_t is replaced by the *tensor* quantity $C_\theta(k/\varepsilon)\rho\langle u_i'u_j'\rangle$. The constant C_θ is generally given a value of about 0.3 (Daly and Harlow 1970). Equation (5.46) can be used *in lieu* of the simpler Eq. (5.6) also in the context of eddy viscosity models such as the k–ε, expressing, of course, the Reynolds stresses $\rho\langle u_i'u_j'\rangle$ as $2\mu_t\langle S_{ij}\rangle$; this makes it possible to model turbulent fluxes that are not aligned with the mean temperature.

5.8 Differential Reynolds Stress/Flux Models

The most rigorous answer to the problems arising from the eddy viscosity assumption (Boussinesq approximation for the turbulent stresses) is to write and solve separate transport equations for each of the six independent components of the turbulent stress tensor $\wp_{t,ij} = -\rho\langle u_i'u_j'\rangle$ (and, in problems involving heat transfer, also for the three components of the turbulent flux vector $q_i = \rho c_p\langle u_i'T'\rangle$). Turbulence models based on this approach are called *Differential Stress Models* (DSM).

A formally exact transport equation for $\rho\langle u_i'u_j'\rangle$ (terms *opposite* to the Reynolds stresses!) can be built starting from the Navier–Stokes Eq. (1.17), substituting $u_i = \langle u_i \rangle + u_i'$ etc., multiplying the i-th equation by u_j' and the j-th by u_i', summing and finally averaging term by term in time. Considering only constant-density fluids, neglecting for simplicity the body force term ρa_i and aacounting for the properties of the time-averaging operator $\langle\langle u_i \rangle \langle u_j \rangle\rangle - \langle u_i \rangle \langle u_j \rangle = 0$, $\langle\langle u_i \rangle u_j'\rangle = \langle u_i'\langle u_j\rangle\rangle = 0$, one has

$$\frac{\partial \rho \langle u_i' u_j' \rangle}{\partial t} + \langle u_k \rangle \frac{\partial \rho \langle u_i' u_j' \rangle}{\partial x_k} = -\left[\rho \langle u_j' u_k' \rangle \frac{\partial \langle u_i \rangle}{\partial x_k} + \rho \langle u_i' u_k' \rangle \frac{\partial \langle u_j \rangle}{\partial x_k} \right] - 2\mu \left\langle \frac{\partial u_i'}{\partial x_k} \frac{\partial u_j'}{\partial x_k} \right\rangle$$

$$+ \left\langle p' \left(\frac{\partial u_i'}{\partial x_j} + \frac{\partial u_j'}{\partial x_i} \right) \right\rangle - \frac{\partial}{\partial x_k} \left[\rho \langle u_i' u_j' u_k' \rangle - \mu \frac{\partial \langle u_i' u_j' \rangle}{\partial x_k} + \langle p' \left(\delta_{jk} u_i' + \delta_{ik} u_j' \right) \rangle \right]$$

$$\tag{5.47}$$

The terms at the RHS represent production by shear, destruction (dissipation), redistribution and diffusion of $\rho\langle u_i'u_j'\rangle$, respectively. Einstein's convention of implicit summation over repeated indices is adopted, and δ_{ij} is Kronecker's delta.

Writing Eq. (5.47) for the normal turbulent stresses ($i = j$) and summing for $i = 1$ to 3, an exact balance equation is obtained for $k = \frac{1}{2}\langle u_i'u_i'\rangle$ (Pope 2000):

$$\frac{\partial \rho k}{\partial t} + \langle u_i\rangle \frac{\partial \rho k}{\partial x_i} = -\rho\left\langle u_i'u_j'\right\rangle \frac{\partial \langle u_i\rangle}{\partial x_j} - \mu\left\langle \frac{\partial u_i'}{\partial x_j}\frac{\partial u_i'}{\partial x_j}\right\rangle$$
$$- \frac{\partial}{\partial x_i}\left[\frac{1}{2}\rho\left\langle u_i'u_j'u_j'\right\rangle - \mu\frac{\partial k}{\partial x_i} + \left\langle p'u_i'\right\rangle\right] \qquad (5.48)$$

The terms at the RHS represent production by shear, dissipation and diffusion of k, respectively. This last term, in its turn, is composed of turbulent dispersion (triple correlation), viscous diffusion and pressure transport (cross-correlation of pressure and velocity fluctuations). The redistribution terms of Eq. (5.47) sum up to zero so that they are not present in the balance of k in Eq. (5.48). Note also that the production term in Eq. (5.48) is consistent with general expression of P_k given in Eq. (5.13).

Figure 5.7 reports radial profiles of the various terms in the budget of turbulent kinetic energy as computed by the author via DNS for fully developed turbulent flow in a circular duct at a friction velocity Reynolds number Re_τ, based on the radius R, of 476 (yielding a bulk Reynolds number $\mathrm{Re}\approx16,000$). The terms in Eq. (5.48) are made dimensionless dividing them by $\rho u_\tau^4/\nu$ (i.e., they are expressed in wall units), and also the distance y from the wall is expressed in wall units as $y^+ = yu_\tau/\nu$.

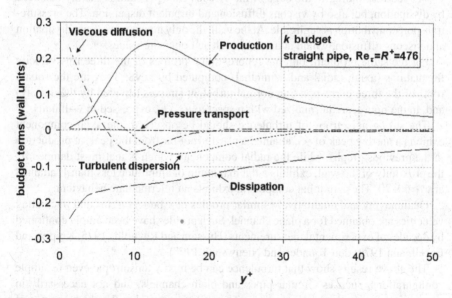

Fig. 5.7 Radial profiles of the different terms in the budget of k in the wall region of a circular duct for fully developed turbulent flow at $\mathrm{Re}_\tau = 476$ ($\mathrm{Re}\approx16,000$) (author's own results)

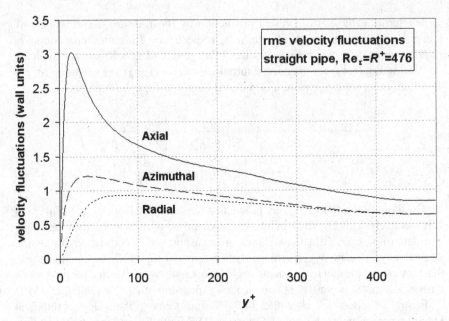

Fig. 5.8 Radial profiles of the three rms velocity fluctuations in a circular duct for fully developed turbulent flow at $Re_\tau = 476$ (Re $\approx$ 16,000) (author's own results)

Note that production by shear exhibits a broad peak at a distance from the wall $y^+\approx12$ (corresponding to the edge of the viscous sublayer) and is mainly balanced by dissipation, but also by viscous diffusion and turbulent dispersion. The pressure-transport contribution is negligible. At the wall, the only nonzero terms are dissipation and viscous diffusion, which attain very high and opposite values.

For the same problem, Fig. 5.8 reports radial profiles of the three rms velocity fluctuations (axial, radial and azimuthal) computed by DNS. They are the square roots of the three corresponding normal turbulent stresses (divided by the density) and, in the figure, are normalized with respect to u_τ, i.e., expressed in wall units.

The strong anisotropy of turbulence should be observed: the axial component exhibits a narrow peak of ~3, attained a $y^+\approx15$ (not far from the peak of production in the previous graph), while the radial component, which is that most damped by the proximity of the wall, exhibits a flat maximum of amplitude less than 1, attained at $y^+\approx60$–70. The azimuthal component exhibits an intermediate behaviour.

Qualitatively and qualitatively similar profiles of k budget terms and fluctuating velocities are obtained for a plane channel. Such profiles have been amply confirmed by decades of experimental measurements (Hussain and Reynolds 1975; Kreplin and Eckelmann 1979; den Toonder and Nieuwstadt 1997).

The above results show that turbulence can be highly anisotropic even in simple configurations, such as circular pipes and plain channels, and not necessarily in geometrically complex configurations. They also suggest that the shortcomings of eddy viscosity models such as the k–ε do *not* arise from turbulence anisotropy as such, since these models yield excellent predictions of the mean velocities (but also of turbulent kinetic energy levels) when applied to the simple flows considered above.

The system of 6 Eqs. (5.47), together with the continuity and Navier–Stokes equations for mean velocities, would allow the mean quantities $\langle u_i \rangle$, $\langle p \rangle$ and the Reynolds stresses $\rho \langle u_i' u_j' \rangle$ to be computed, were it not for the (unavoidable) presence of terms containing *new* unknowns. These terms must be approximated as functions of mean quantities in order to close the problem. Among many possible ways of doing this, the approach that will be described here is basically that proposed by Hanjaliç and Launder (1972) and Launder et al. (1975).

(a)   *Destruction term*: assuming the dissipative scales of turbulence to be isotropic, this can be modelled as

$$2\mu \left\langle \frac{\partial u_i'}{\partial x_k} \frac{\partial u_j'}{\partial x_k} \right\rangle = \frac{2}{3} \delta_{ij} \rho \varepsilon \qquad (5.49)$$

Equation (5.49) states that the destruction terms act only on the *normal* Reynolds stresses, are equal in the three directions, and add up to twice the dissipation ε (multiplied by the density ρ). The factor 2 is necessary because the sum of the three normal Reynolds stresses is equal to twice the turbulent kinetic energy k (multiplied by the density ρ). The presence of ε requires a further transport equation for this quantity, modelled after Eq. (5.17) but suitably modified to explicitly accommodate for the six Reynolds stresses (see below).

(b)   *Diffusion terms*: the correlation of fluctuating pressure and velocities is usually small (see Fig. 5.6) and can be neglected. The term proportional to the molecular viscosity μ contains only Reynolds stresses and thus does not require a closure model. The only term to be modelled is the divergence of the triple correlation $\langle u_i' u_j' u_k' \rangle$ (turbulent dispersion term). The authors show that an acceptable approximation is

$$-\left\langle u_i' u_j' u_k' \right\rangle = c_t \frac{k}{\varepsilon} \left[\left\langle u_i' u_l' \right\rangle \frac{\partial \left\langle u_j' u_k' \right\rangle}{\partial x_l} + \left\langle u_j' u_l' \right\rangle \frac{\partial \left\langle u_k' u_i' \right\rangle}{\partial x_l} + \left\langle u_k' u_l' \right\rangle \frac{\partial \left\langle u_i' u_j' \right\rangle}{\partial x_l} \right]$$

$$(5.50)$$

in which c_t is a constant and the multiplier k/ε is the characteristic time scale of the turbulence structures responsible for the turbulent dispersion.

(c)   *Redistribution term*: as Eq. (5.47) shows, this arises from the correlation between the fluctuating components of pressure and velocity gradients. The name is due to the fact that this term redistributes turbulence energy among the three normal turbulent stresses, thus reducing the difference between their amplitudes. On the basis of pioneering studies by Chou (1945) and Rotta

(1951), Hanjalić and Launder express this term as the sum of three contributions $(\Phi_{ij})_1, (\Phi_{ij})_2, (\Phi_{ij})_w$.

The contribution $(\Phi_{ij})_1$ expresses the mutual interaction between fluctuating velocity components and can be assumed to be proportional to the deviatoric part of the turbulent stress tensor:

$$(\Phi_{ij})_1 = -C_{\Phi 1}\rho\frac{\varepsilon}{k}\left(u_i'u_j' - \frac{2}{3}\delta_{ij}k\right) \tag{5.51}$$

in which $C_{\Phi 1}$ is a model constant of the order of 2.5–2.8.

The contribution $(\Phi_{ij})_2$ expresses the interaction between fluctuating velocity components and mean velocity gradients. Its approximate expression according to Hanjalić and Launder is

$$(\Phi_{ij})_2 = -\gamma_t\left(P_{ij} - \frac{2}{3}\delta_{ij}P_k\right) \tag{5.52}$$

in which γ_t is a model constant of the order of 0.6, P_{ij} is the production of $\rho\langle u_i'u_j'\rangle$, first term at the RHS of Eq. (5.47), and P_k is the production rate of ρk (note that $P_k = P_{jj}/2$). More complex expressions for $(\Phi_{ij})_2$ have also been proposed, based on contracting mean velocity gradients with a fourth-order tensor, but they will not be discussed here.

Finally, the contribution $(\Phi_{ij})_w$ is the so called "wall reflection term", which models the contribution of solid walls to the anisotropy of normal turbulent stresses. A possible approximation is

$$(\Phi_{ij})_w = \left[C_{1w}\rho\frac{\varepsilon}{k}\left(\langle u_i'u_j'\rangle - \frac{2}{3}\delta_{ij}k\right) + C_{2w}\left(P_{ij} - B_{ij}\right)\right]\frac{k^{3/2}}{\varepsilon y} \tag{5.53}$$

in which $C_{1w} = 0.125$ and $C_{2w} = 0.015$ are two further model constants, y is the distance from the nearest wall, P_{ij} was defined above, and B_{ij} is defined as

$$B_{ij} = -\left[\rho\langle u_j'u_k'\rangle\frac{\partial\langle u_k\rangle}{\partial x_i} + \rho\langle u_i'u_k'\rangle\frac{\partial\langle u_k\rangle}{\partial x_j}\right] \tag{5.54}$$

Also for the wall reflection terms a large variety of alternative and more complex models have been proposed in the literature.

Taking account of the various approximations described above, the transport Eq. (5.47) for $\rho\langle u_i'u_j'\rangle$ can be written in closed form as

$$\frac{\partial\rho\langle u_i'u_j'\rangle}{\partial t} + \langle u_k\rangle\frac{\partial\rho\langle u_i'u_j'\rangle}{\partial x_k} = -\left[\rho\langle u_j'u_k'\rangle\frac{\partial\langle u_i\rangle}{\partial x_k} + \rho\langle u_i'u_k'\rangle\frac{\partial\langle u_j\rangle}{\partial x_k}\right] - \frac{2}{3}\delta_{ij}\rho\varepsilon$$

$$- C_{\Phi 1}\rho\frac{\varepsilon}{k}\left(\overline{u_i'u_j'} - \frac{2}{3}\delta_{ij}k\right) - \gamma\left(P_{ij} - \frac{2}{3}\delta_{ij}P\right)$$

$$+ \left[C_{1w}\rho\frac{\varepsilon}{k}\left(\langle u_i'u_j'\rangle - \frac{2}{3}\delta_{ij}k\right) + C_{2w}\left(P_{ij} - B_{ij}\right)\right]\frac{k^{3/2}}{\varepsilon y}$$

$$+ \frac{\partial}{\partial x_k}\mu\frac{\partial\langle u_i'u_j'\rangle}{\partial x_k} - c_s\frac{\partial}{\partial x_k}\frac{\rho k}{\varepsilon}$$

$$\times \left[\langle u_i'u_l'\rangle\frac{\partial\langle u_j'u_k'\rangle}{\partial x_l} + \langle u_j'u_l'\rangle\frac{\partial\langle u_k'u_i'\rangle}{\partial x_l} + \langle u_k'u_l'\rangle\frac{\partial\langle u_i'u_j'\rangle}{\partial x_l}\right]$$

$$(5.55)$$

This set of 6 independent equations includes several constants ($C_{\Phi 1}$, C_{1w}, C_{2w}, γ_t, c_t) whose values must be optimized by comparison with experimental, analytical or asymptotic results. In Eq. (5.55), besides the turbulent (Reynolds) stresses $\rho\langle u_i'u_j'\rangle$, also the quantities k and ε are present. Of course, k is just a shorthand for $\frac{1}{2}\langle u_i'u_i'\rangle$ and thus does not introduce new unknowns. In regard to the dissipation ε, as anticipated above it requires a separate transport equation, independent from Eqs. (5.55) and modelled after Eq. (5.17). A possible form of the ε equation is

$$\frac{\partial\rho\varepsilon}{\partial t} + \frac{\partial\rho\langle u_i\rangle\varepsilon}{\partial x_i} = -C_{\varepsilon 1}\frac{\varepsilon}{k}\rho\langle u_i'u_j'\rangle\frac{\partial\langle u_i\rangle}{\partial x_j} + C_\varepsilon\frac{\partial}{\partial x_k}\left(\rho\langle u_k'u_j'\rangle\frac{k}{\varepsilon}\frac{\partial\varepsilon}{\partial x_j}\right) - C_{\varepsilon 2}\rho\frac{\varepsilon^2}{k}$$

$$(5.56)$$

The reader will recognize similarities and differences between this last equation, formulated in the context of Reynolds stress transport models, and Eq. (5.17), formulated in the context of eddy viscosity models. In particular, the first term at the RHS of Eq. (5.56) represents the diffusion of ε, written by a Generalized Gradient Diffusion Hypothesis (GGDH) similar to that for turbulent scalar fluxes in Eq. (5.46).

Like Eq. (5.17), which required the three model constants σ_ε, C_1 and C_2, also Eq. (5.56) requires three further model constants $C_{\varepsilon 1}$, $C_{\varepsilon 2}$, C_ε for which Hanjalič and Launder suggest the values $C_{\varepsilon 1} = 1.44$, $C_{\varepsilon 2} = 1.9$, $C_\varepsilon = 0.15$.

In regard to the transport of enthalpy and other scalars, also in differential stress models, as in algebraic ones, the gradient diffusion hypothesis can be consistently replaced by the Generalized Gradient Diffusion Hypothesis, Eq. (5.44).

Reynolds stress transport models, or *second order models*, imply very complex expressions and require a large number of model constants to be tuned. Compared with eddy viscosity two-equation models, in three-dimensional flows they require the solution of the seven transport Eqs. (5.55), (5.56) instead of two and are much more demanding in terms of computational resources.

Also, since the purely diffusive terms of eddy viscosity models are replaced by *non-diffusive* terms, these models may exhibit numerical stability problems. Often, simulations using Reynolds stress models are "primed" by a certain number of iterations or time steps based on an eddy viscosity model to help convergence.

On the other hand, as anticipated in introducing them in Sect. 5.7, second order models are able correctly to predict flow quantities affected by the anisotropy of the normal Reynolds stresses, or by the misalignment of the Reynolds stress tensor with the mean strain rate tensor. Examples include the reattachment length in separated flows, the spreading rate of submerged round or planar jets, the secondary recirculation in non-circular ducts and the velocity field in rapidly swirling flows.

An example of this last issue is given in Fig. 5.9. It regards the finite volume simulation of the turbulent flow field in a mechanically stirred cylindrical vessel of the so called *unbaffled* type, i.e. devoid of the baffles which are commonly mounted on the peripheral wall to promote mixing (Ciofalo et al. 1996).

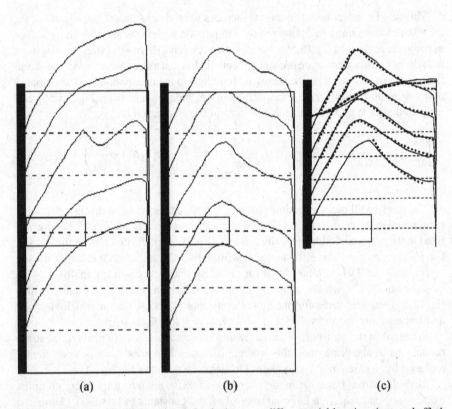

(a)   (b)   (c)

Fig. 5.9 Radial profiles of the azimuthal velocity u_ϕ at different axial locations in an unbaffled vessel stirred by a paddle impeller. **a** k–ε model, vessel closed by a top lid; **b** Reynolds stress model, same geometry; **c** Reynolds stress model predictions (solid lines) and experimental results (symbols), open vessel. Experimental and computed profiles of the free surface vortex are also reported. From results of Ciofalo et al. (1996)

For this problem, simulations were conducted in the rotating reference frame of the impeller in order to avoid the problems raised by the motion of the impeller blades. The cylindrical walls were described as sliding walls by suitably modifying the corresponding no slip boundary conditions. Appropriate source terms were added to the right hand side of the momentum equations to account for the centrifugal and Coriolis inertial forces that arise in the non-inertial (rotating) reference frame.

Radial profiles of the mean azimuthal velocity u_ϕ at different axial locations, computed by the k–ε model and by a Reynolds stress model for the case in which the vessel top is *closed* by a lid, are reported in graphs (a) and (b), respectively. The k–ε predictions resemble a rigid-body rotation, while the second-order model predicts non-monotonic velocity profiles, which are empirically known to be correct.

Graph (c) compares Reynolds stress model predictions with experimental results, taken from the classic book of Nagata (1975), for a similar but superiorly *open* vessel, in which the free surface forms a typical central vortex; the agreement is excellent not only qualitatively but also quantitatively. The graph shows also the experimental and predicted shapes of the free surface of the central vortex and confirms a good agreement also for this quantity.

References

AEA Technology (1994) CFDS-FLOW3D: user guide. In: Computational fluid dynamics services. Harwell Laboratories, UK

Amano RS, Jensen MK, Goel P (1983) A numerical and experimental investigation of turbulent heat transport downstream from an abrupt pipe expansion. ASME J Heat Transf 105:862–869

Arpaci VS, Larsen PS (1984) Convection Heat Transfer. Prentice-Hall, Englewood Cliffs, NJ

Baldwin BS, Lomax H (1978) Thin-layer approximation and algebraic model for separated turbulent flows. AIAA Paper 78–257

Burns AD, Jones IL, Kightley JR, Wilkes NS (1989) Harwell-FLOW3D Release 2 User Manual. UKAEA Report AERE-R (Draft), July

Cebeci T, Smith AMO (1974) Analysis of turbulent boundary layers. Academic Press, New York

Chen Y-S, Kim S-W (1987) Computation of turbulent flows using an extended k-epsilon turbulence closure model. NASA Technical Report NASA-CR-179204

Chieng CC, Launder BE (1980) On the calculation of turbulent heat transport downstream from an abrupt pipe expansion. Numer Heat Transf 3:189–207

Chima RV, Giel PW, Boyle RJ (1993) Algebraic turbulence model for three-dimensional viscous flow. In: Rodi W, Martelli F (eds) Engineering turbulence modelling and experiments 2. Elsevier, Amsterdam, pp 775–784

Chou PY (1945) On velocity correlations and the solution of the equations of turbulent fluctuation. Quart Appl Math 3(1):38–54

Ciofalo M, Collins MW (1989) k-ε predictions of heat transfer in turbulent recirculating flows using an improved wall treatment. Numer Heat Transf B 15:21–47

Ciofalo M, Palagonia M (1996) Turbulent flow separation past triangular obstacles. Quaderno No. 10/96, Dipartimento di Ingegneria Nucleare, Università di Palermo

Ciofalo M, Brucato A, Grisafi F, Torraca N (1996) Turbulent fluid flow in closed- and free-surface unbaffled tanks stirred by radial impellers. Chem Eng Sci 51:3557–3573

Collins MW, Ciofalo M, Di Piazza I (1998) Filtering of the Navier-Stokes equations in the context of time-dependent flows. In: Rahman M, Comini G, Brebbia CA (eds) Advances in Fluid Mechanics 2. Computational Mechanics Publications, Southampton, pp 69–81

Cruz DOA, Silva Freire AP (1998) On single limits and the asymptotic behaviour of separating turbulent boundary layers. Int J Heat Mass Transf 41:2097–2112

Daly BJ, Harlow FH (1970) Transport equations in turbulence. Phys Fluids 13:2634–2649

den Toonder JMJ, Nieuwstadt FTM (1997) Reynolds number effects in a turbulent pipe flow for low to moderate Re. Phys Fluids 9:3398–3409

Di Piazza I, Ciofalo M (2010) Numerical prediction of turbulent flow and heat transfer in helically coiled pipes. Int J Thermal Sci 49:653–663

Hanjalić K, Launder BE (1972) A Reynolds stress model of turbulence and its application to thin shear flows. J Fluid Mech 52:609–638

Hanjalić K, Launder BE (1980) Sensitizing the dissipation equation to irrotational strains. ASME J Fluids Eng 102:34–40

Hinze JO (1975) Turbulence, 2nd edn. McGraw-Hill, New York

Hussain AKMF, Reynolds WC (1975) Measurements in fully developed turbulent channel flow. ASME J Fluids Eng 97:568–578

Jayatilleke CLV (1969) The influence of Prandtl number and surface roughness on the resistance of the laminar sublayer to momentum and heat transfer. Progr Heat Mass Transf 1:193–329

Kolmogorov AN (1942) Equations of turbulent motion of an incompressible fluid. Izv Acad Nauk USSR Phys 6:56–58

Kreplin H-P, Eckelmann H (1979) Behavior of the three fluctuating velocity components in the wall region of a turbulent channel flow. Phys Fluids 22:1233–1239

Lam CKG, Bremhorst KA (1981) Modified form of the k–ε model for predicting wall turbulence. ASME J Fluids Eng 103:456–460

Landau LD, Lifshitz EM (1959) Fluid Mechanics. Pergamon Press, Reading, MA

Launder BE, Spalding DB (1972) Mathematical Models of Turbulence. Academic Press, London

Launder BE, Spalding DB (1974) The numerical computation of turbulent flows. Comp Meth Appl Mech Eng 3:269–289

Launder BE, Sharma BI (1974) Application of the energy–dissipation model of turbulence to the calculation of flow near a spinning disc. Lett Heat Mass Transf 1:131–138

Launder BE, Reece GJ, Rodi W (1975) Progress in the development of a Reynolds-stress turbulence closure. J Fluid Mech 68:537–566

Menter FR (1993) Zonal two-equations k–ω turbulence models for aerodynamic flows. AIAA Paper 93–2906

Menter FR (1994) Two-equation eddy-viscosity turbulence models for engineering applications. AIAA J 32:269–289

Menter FR, Kuntz M, Langtry R (2003) Ten years of industrial experience with the SST turbulence model. In: Hanjalić K, Nagano Y, Tummers M (eds) Turbulence, heat and mass transfer 4. Begell House, Inc

Mohammadi B, Pironneau O (1994) Analysis of the k-epsilon turbulence model. Wiley, New York

Nagano Y, Hishida M (1987) Improved form of the k–ε model for wall turbulent shear flows. ASME J Fluids Eng 109:156–160

Nagata S (1975) Mixing: principles and applications. Wiley, New York

Patel VC, Rodi W, Scheuerer G (1985) Turbulence models for near–wall and low–Reynolds number flows: a review. AIAA J 23(9):1308–1319

Pope SB (2000) Turbulent flows. Cambridge University Press, Cambridge, UK

Rotta JC (1951) Statistische Theorie Nichthomogener Turbulenz. Z Phys 129:547–572

Schlichting H (1968) Boundary-layer theory. Pergamon Press, London

Spalart PR, Allmaras SR (1992) A one-equation turbulence model for aerodynamic flows. AIAA Paper 92-0439

Speziale CG (1987) On nonlinear k-l and k-ε models of turbulence. J Fluid Mech 178:459–475

Tanda G, Ciofalo M, Stasiek JA, Collins MW (1995) Experimental and numerical study of forced convection heat transfer in a rib-roughened channel. In: Proceedings 13th National conference of UIT (Unione Italiana di Termofluidodinamica), Bologna, 22–23 June 1995, p 243–254

Thomas CE, Morgan K, Taylor C (1981) A finite element analysis of flow over a backward facing step. Comput Fluids 9:265–278

Vogel JC, Eaton JK (1985) Combined heat transfer and fluid dynamics measurements downstream of a backward-facing step. ASME J Heat Transf 107:922–929

Wilcox DC (1988) Reassessment of the scale-determining equation for advanced turbulence models. AIAA J 26:1299–1310

Yakhot V, Orszag SA (1986) Renormalization group analysis of turbulence. I. Basic Theory. J Sci Comp 1:1–51

Yakhot V, Orszag SA, Thangam S, Gatski TB, Speziale CG (1992) Development of turbulence models for shear flows by a double expansion technique. Phys Fluids A 4:1510–1520

Chapter 6
Turbulence in Natural and Mixed Convection

Reality is that which when you stop believing in it, it doesn't go away
Philip K. Dick, *V.A.L.I.S.*

Abstract Turbulence can be caused or affected by density gradients, notably associated with thermal stratification. The interaction of density gradients with turbulence is discussed, the relevant additional source terms in the transport equations for Reynolds stresses or turbulent kinetic energy and dissipation are derived, and their influence on flow and temperature field is analysed with the aid of examples from the literature. The last section provides some information on turbulence in the atmospheric boundary layer.

Keywords Natural convection · Rayleigh–Taylor instability · Buoyancy · Thermal stratification · Richardson number · Monin–Obukhov length

6.1 Buoyancy and Turbulence

As mentioned in Sect. 1.5, *natural convection* (buoyancy) occurs in the simultaneous presence of:

- a body force (e.g. gravity or the centrifugal and Coriolis forces arising in non-inertial reference frames), which will appear as a source term $f_{V,i} = \rho a_i$ (force per unit volume) at the RHS of the i-th momentum equation;
- a density gradient $\partial \rho / \partial x_i$ with nonzero component in the direction of the said body force ($f_{V,i} \, \partial \rho / \partial x_i \neq 0$).

If forced convection is also present, the resulting condition is called *mixed convection*. The focus here is how buoyancy produces, suppresses or modifies turbulence, and how these effects are captured by existing turbulence models.

From the *physical* point of view, buoyancy affects turbulence both indirectly and directly:

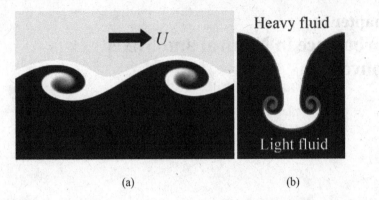

Fig. 6.1 Typical instability mechanisms that can produce turbulence. **a** Kelvin–Helmholtz insta-bility, characteristic of boundary layer flows; **b** Rayleigh–Taylor instability, characteristic of density stratifications

- indirectly, via the modification (or the onset itself) of a mean motion which, in its turn, produces or suppresses turbulence by the mechanisms discussed in the above sections, generally mediated by shear stresses;
- directly, via specific production or suppression mechanisms based on the estab-lishment of regions with either unstable ($f_{V,i} \, \partial \rho / \partial x_i < 0$) or stable ($f_{V,i} \, \partial \rho / \partial x_i > 0$) density stratification.

In Fig. 6.1, the two mechanisms are respectively exemplified by the two instabil-ities of Kelvin–Helmholtz (a), typical of boundary layer flows, and Rayleigh–Taylor (b), typical of density stratifications. These two mechanisms are not the only ones that can produce turbulence, but they were chosen here as representative of the indirect and direct effects of buoyancy.

From the point of view of *mathematical modelling*, a distinction has to be made between different modelling approaches.

- In Direct Numerical Simulation (DNS), no special modelling is required to account for either of the above mechanisms, which will automatically emerge from the time-dependent solution of the governing equations.
- In Reynolds Averaged Navier Stokes (RANS) models, *indirect* effects are implicitly accounted for in modelling turbulence production by *shear*. On the contrary, the *direct* effects of buoyancy on turbulence require a specific modelling because they do not depend on the *mean* density distribution, but rather on the *instantaneous* one, unresolved in RANS turbulence models.
- In Large-Eddy Simulation (LES), the situation is intermediate; indirect effects are implicitly accounted for as in RANS models, but direct effects are implicitly accounted for only at the resolved scales, while at subgrid scales they should be explicitly modelled. Only few subgrid models do so; they will not be discussed here for the sake of brevity.

6.2 Direct Contribution of Buoyancy to the Balance of Reynolds Stresses and k

The exact Reynolds stress transport Eq. (5.47) were obtained from the Navier–Stokes Eq. (1.17) neglecting the body forces ρa_i in these latter. If these terms are included, it is easy to demonstrate that the production term P_{ij} in Eq. (5.47) is augmented by the term

$$G_{ij} = \left\langle u_i' f_{V,j}' \right\rangle + \left\langle u_j' f_{V,i}' \right\rangle \tag{6.1}$$

in which $f_{V,i}'$ is the fluctuating component of the body force per unit volume $f_{V,i} = \rho a_i$.

For $a_i = $ constant, $f_{V,i}'$ arises solely from the density fluctuations ρ' so that

$$f_{V,i}' = a_i \rho' \tag{6.2}$$

Therefore, in this case the production term G_{ij} in Eq. (6.1) can be written

$$G_{ij} = a_i \left\langle u_j' \rho' \right\rangle + a_j \left\langle u_i' \rho' \right\rangle \tag{6.3}$$

Note that the assumption $a_i = $ constant, on which Eq. (6.3) is based, is trivially true for the gravity acceleration and is still true, provided the angular drag velocity does not change in time, for the centrifugal acceleration. However, it is false, in general, for the Coriolis acceleration, which is a function of the velocity relative to a rotating frame and thus exhibits fluctuations in a turbulent flow. Therefore, this last case (which includes large-scale atmospheric turbulence) requires a different and more complex treatment and will not be discussed here.

Density fluctuations can be due to fluctuations of temperature, concentration (e.g. salinity) and, in compressible fluids, pressure. In the following, we will assume that they arise from *temperature* fluctuations. The extension to the case of concentration fluctuations (or simultaneous temperature and concentration fluctuations) is straight-forward, while turbulent natural convection in compressible fluids requires a different conceptual apparatus (Gatski and Bonnet 2013).

According to the Boussinesq approximation for thermal buoyancy terms, introduced in Sect. 1.5, one has $\rho' = -\beta_T \rho T'$, $\beta_T = -(1/\rho)(\partial \rho / \partial T)$ being the fluid's cubic dilatation coefficient. Therefore, the velocity-density correlations can be expressed as

$$\left\langle u_j' \rho' \right\rangle = -\beta_T \rho \left\langle u_j' T' \right\rangle \tag{6.4}$$

and the buoyancy production term G_{ij} in Eq. (6.3) becomes

$$G_{ij} = -\beta_T \rho \left(a_i \left\langle u_j' T' \right\rangle + a_j \left\langle u_i' T' \right\rangle \right) \tag{6.5}$$

which, making use of Eq. (5.4) for the turbulent fluxes, $q_{t,i} = \rho c_p \langle u_i' T' \rangle$, can be written

$$G_{ij} = -\frac{\beta_T}{c_p} \left(a_i q_{t,j} + a_j q_{t,i} \right) \tag{6.6}$$

The complete and exact transport equation for the generic Reynolds stress, accounting for production both by shear and by buoyancy, is a modified form of Eq. (5.47) obtained by adding to the RHS the terms G_{ij} expressed by Eq. (6.6). Its validity rests on that of the assumption $g_i = $ constant (with respect to time) and of the Boussinesq approximation.

Turbulent fluxes can be modelled by the Generalized Gradient Diffusion Hypothesis (GGDH), Eq. (5.46), so that the production terms in Eq. (6.6) become

$$G_{ij} = C_\theta \rho \frac{k}{\varepsilon} \beta_T \left(a_i \cdot \left\langle u_j' u_k' \right\rangle + a_j \cdot \left\langle u_i' u_k' \right\rangle \right) \frac{\partial \langle T \rangle}{\partial x_k} \tag{6.7}$$

with implicit summation over the index k and $C_\theta \approx 0.3$ (Daly and Harlow 1970).

Within the context of eddy viscosity models, including the common two-equation $k - \varepsilon$ and $k - \omega$ models, what is required is a term G_k expressing the extra production of turbulent kinetic energy due to buoyancy. An exact expression for this term is obtained by observing that, since $\rho k = \tau_{jj}/2$ (implicit summation on j), then $G_k = G_{jj}/2$; using Eq. (6.3) for the terms G_{jj}, one has

$$G_k = a_j \left\langle u_j' \rho' \right\rangle \tag{6.8}$$

Adopting, as above, the Boussinesq approximation $\rho' = -\rho \beta_T T'$ for density fluctuations, the term G_k can be written

$$G_k = -\rho \beta_T a_j \left\langle u_j' T' \right\rangle \tag{6.9}$$

(with implicit summation on j); or, remembering the definition of the turbulent heat fluxes as $q_{t,j} = \rho c_p \langle u_i' T' \rangle$:

$$G_k = -\frac{\beta_T}{c_p} a_j q_{t,j} \tag{6.10}$$

In regard to a closure equation for the turbulent fluxes $q_{t,j}$, in the context of eddy viscosity models the Simple Gradient Diffusion Hypothesis (SGDH) of Eq. (5.3) can consistently be used. Equation (6.9) then yields the closed-form expression

$$G_k = \beta_T \Gamma_t a_j \frac{\partial \langle T \rangle}{\partial x_j} \tag{6.11}$$

in which $\Gamma_t = \mu_t/\sigma_t$ (σ_t being the turbulent Prandtl number).

If the only acceleration is that due to gravity and the axis $x_3 = z$ is directed upward, then $a_3 = -g$, $a_1 = a_2 = 0$ and Eq. (6.11) reduces to

$$G_k = -\beta_T \Gamma_t g \frac{\partial \langle T \rangle}{\partial z} \Gamma \tag{6.12}$$

In this case, G_k is positive (net production) if $\partial \langle T \rangle / \partial z < 0$, i.e. if the temperature decreases with the height (unstable stratification), whereas G_k is negative (net suppression) if $\partial \langle T \rangle / \partial z > 0$, i.e. if the temperature increases with the height (stable stratification). An unstable stratification promotes turbulence, whereas a stable stratification tends to suppress turbulence (if present).

In mixed convection, the relative importance of buoyancy and forced convection is measured by the *Richardson number*, defined as

$$Ri = = \frac{-G_k}{P_k} \tag{6.13}$$

i.e. as the opposite of the ratio between buoyancy and shear production terms.

P_k is generally positive (save in the presence of spatial accelerations such as occur in converging ducts, which tend to suppress turbulence); therefore, by its definition in Eq. (6.13), Ri is positive when G_k is negative, i.e. when buoyancy tends to suppress turbulence (stable stratification), whereas it is negative when both shear and buoyancy act together to promote turbulence. For Ri > 0.25 the flow is considered stable (the suppression of turbulence due to a stable stratification overcomes the production due to shear), while for Ri < 0.25 the flow is considered unstable (turbulent).

If one wishes to avoid the gradient-diffusion assumption, then G_k must be written as in Eq. (6.9) and P_k as $\tau_{t,ij} \langle S_{ij} \rangle$, which, in constant-density fluids, becomes

$$P_k = -\rho \langle u_j' u_k' \rangle \frac{\partial \langle u_j \rangle}{\partial x_k} \tag{6.14}$$

The resulting expression for Ri is

$$Ri_f = \frac{\rho \beta_T a_j \langle u_j' T' \rangle}{-\rho \langle u_j' u_k' \rangle \frac{\partial \langle u_j \rangle}{\partial x_k}} \tag{6.15}$$

which is called *flux Richardson number* (Ri_f).

If, on the contrary, the simple gradient diffusion hypothesis (SGDH) is accepted both for P_k and for G_k, then G_k must be written as in Eq. (6.11) and P_k as

$$P_k = \mu_t \left(\frac{\partial \langle u_j \rangle}{\partial x_k} + \frac{\partial \langle u_k \rangle}{\partial x_j} \right) \frac{\partial \langle u_j \rangle}{\partial x_k}, \tag{6.16}$$

In this case, taking account that $\Gamma_t = \mu_t/\sigma_t$, Ri takes the form

$$\text{Ri}_g = = \frac{\beta_T}{\sigma_t} \frac{-a_j \frac{\partial \langle T \rangle}{\partial x_j}}{\left(\frac{\partial \langle u_j \rangle}{\partial x_k} + \frac{\partial \langle u_k \rangle}{\partial x_j} \right) \frac{\partial \langle u_j \rangle}{\partial x_k}} \tag{6.17}$$

which is called *gradient Richardson number* (Ri_g).

6.3 Direct Contribution of Buoyancy to the Balance of Dissipation

Both in Reynolds Stress transport and in eddy viscosity models, buoyancy affects also the balance of dissipation ε or quantities related to it (ω, l, etc.) and its direct effects have to be modelled in the corresponding transport equations.

Limiting the discussion to the $k - \varepsilon$ model, the approach followed by most authors is that of replacing the shear production of ε in Eq. (5.18), i.e. $C_1(\varepsilon/k)P$, with the term

$$C_1 \frac{\varepsilon}{k} (P_k + C_3 G_k) \tag{6.18}$$

in which P_k and G_k are expressed as in the transport equation for k, e.g. Equations (6.16) and (6.11), while C_3 is a further model constant.

The agreement between different authors, however, ends here, and different expressions have been proposed for C_3:

- $C_3 = $ constant, with values ranging from 0.5 to 1;
- $C_3 = $ function of the Richardson number $\text{Ri} = -G_k/P_k$ (Rodi 1987; Yan and Holmstedt 1999; Van Maele and Mercy 2006);
- $C_3 = \tanh(|w/u|)$, in which w e u are the vertical and horizontal components of velocity, so that C_3 tends to 1 if the flow is vertical and directed upward, tends to -1 if the flow is vertical but is directed downward, and vanishes if the flow is horizontal (Lam and Bremhorst 1981; Abe et al. 1994);
- $C_3 = 0$ in a stable stratification ($G_k < 0$) while $C_3 = 1$ in an unstable stratification ($G_k > 0$) (Burns et al. 1989). This last option avoids the possibility that in stably stratified flows the production of ε becomes negative like that of k, possibly causing the variable ε itself to become (unphysically) negative and k to blow up.

Figures 6.2 and 6.3 illustrate a comparative test performed by the author based on DNS results for plane channel flows, due to Nobile et al. (2000) for unstable stratification (Fig. 6.2) and to Lida et al. (2002) for stable stratification (Fig. 6.3).

The graphs compare cross-stream profiles of mean longitudinal velocity (a) and temperature (b) (in wall units) as predicted by DNS and as computed by using the $k - \varepsilon$ model under three alternative assumptions:

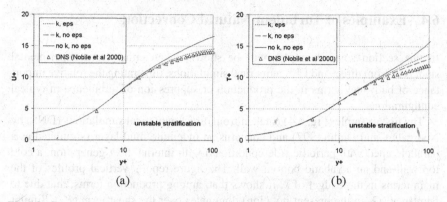

Fig. 6.2 Cross-stream profiles of mean longitudinal velocity (**a**) and temperature (**b**), in wall units, for turbulent flow in a plane channel with *unstable* thermal stratification. Symbols: DNS (Nobile et al. 2000); lines: $k - \varepsilon$ model with the buoyancy term G_k included or not in the equations for k and ε

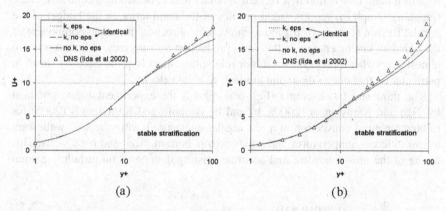

Fig. 6.3 Cross-stream profiles of mean longitudinal velocity (**a**) and temperature (**b**), in wall units, for turbulent flow in a plane channel with *stable* thermal stratification. Symbols: DNS (Lida et al. 2002); lines: $k - \varepsilon$ model with the buoyancy term G_k included or not in the equations for k and ε

1. buoyancy term G_k omitted in the transport equations of both k and ε;
2. buoyancy term G_k present in the transport equation of k but not in that of ε;
3. buoyancy term G_k present both in the transport equation of k and in that of ε, in which it is modelled according to the third option ($C_3 = 0$ if $G_k < 0$, $C_3 = 1$ if $G_k > 0$).

Assumption (c) reduces to (b) (buoyancy term completely omitted in the equation for ε) in the stable stratification case of Fig. 6.3 ($G_k < 0$).

The comparison with DNS results clearly shows that the best option is (b): including the buoyancy term in the transport equation for k improves the accuracy of the predictions, but including it also in the equation for ε is detrimental in the presence of both a stable and an unstable stratification.

6.4   Examples of Turbulent Natural Convection

In this section some examples will be shown, taken either from experimental studies or from direct and LES numerical simulations, illustrating the relative importance of buoyancy terms in the production or suppression of turbulence in typical configurations.

The first example (Fig. 6.4) is taken from direct numerical simulations (DNS) by Nourgaliev and Dinh (1997) and concerns an indefinite fluid layer (simulated as a parallelepiped with periodic side conditions) with internal heat generation, a cold top wall and an adiabatic bottom wall. The figure reports vertical profiles of the main terms in the budget of k. It shows that, among production terms, that due to buoyancy (G_k in the present notation) dominates over the shear term (P_k). Similar results are obtained in Rayleigh-Bènard convection, in which a fluid layer is kept unstably stratified between a lower hot wall and a colder top wall.

The second example (Fig. 6.5) is taken from a LES by Barhaghi et al. (2006) and regards natural convection in a vertical annular channel heated by a central hot tube (graph a). Graph (b) reports radial profiles of the main terms in the budget of the normal Reynolds stress $\langle v_z' v_z' \rangle$ along the vertical direction, principal component of the turbulent kinetic energy. In this case, production by buoyancy, although positive almost everywhere, plays only a minor role with respect to production by shear, in particular by that due to the radial gradient of axial velocity.

The third and last example (Fig. 6.6) regards the experimental data obtained by Tian and Karayiannis (2000a, b) and by Ampofo and Karayiannis (2003) for turbulent natural convection in a rectangular enclosure having the side walls kept at two different temperatures and adiabatic top, bottom, front and back walls. This is one of the most complete and accurate existing data bases on turbulent natural

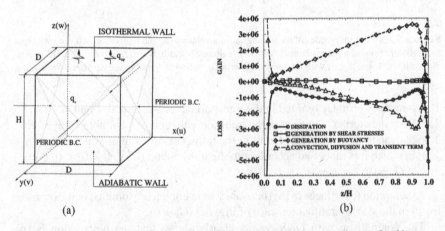

(a)   (b)

Fig. 6.4 DNS of natural convection in a fluid layer with internal heat generation confined between indefinite plane parallel walls. **a** Schematic representation of the problem; **b** vertical profiles of the main terms in the budget of k. Reprinted from Nourgaliev and Dinh (1997) with permission from Elsevier

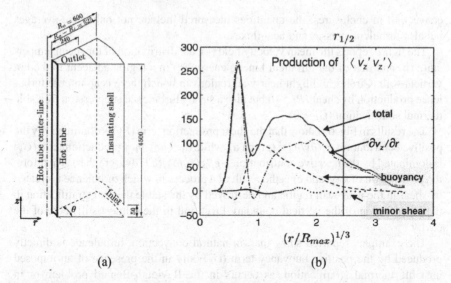

(a)   (b)

Fig. 6.5 LES of natural convection along a hot tube. **a** Schematic representation of the problem; **b** radial profile of the main terms in the budget of $\langle v_z' v_z' \rangle$ (axial normal Reynolds stress). Reprinted from Barhaghi et al. (2006) with permission from Elsevier

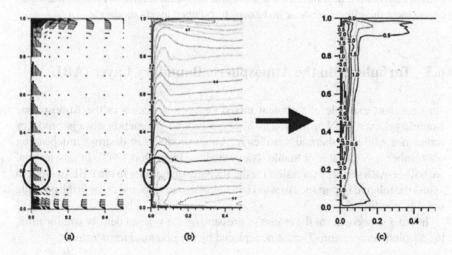

(a)   (b)   (c)

Fig. 6.6 Experimental distributions of mean velocity (**a**), mean temperature (**b**) and turbulent kinetic energy (**c**) near the hot wall of a differentially heated enclosure. Reprinted from Tian and Karayiannis (2000a, b) with permission from Elsevier

convection in enclosures; the quantities measured include not only time-averages but also turbulent stresses and heat fluxes.

The figure reports the mean velocity field (a), the distribution of the mean temperature (b) and that of the turbulent kinetic energy (c) in a region adjacent to the hot vertical wall. Circles highlight near-wall regions in which there is an intense turbulence production by shear ($P_k > 0$) but also a strong turbulence suppression by stable thermal stratification ($G_k < 0$).

The results in Fig. 6.6 show that the shear production term (P_k) is dominated by the positive contribution $\mu_t(\partial v/\partial x)^2$ (graph a), while the *buoyancy* production term (G_k) is dominated by the negative contribution $-g\beta_T(\mu_t/\sigma_t)(\partial T/\partial y)$ (graph b). Therefore, the distribution of k (graph c) is the result of a process in which turbulence is created by shear in the near-wall region and destroyed by the stable thermal stratification in the outer region of the vertical boundary layer and in the (almost still) core of the cavity.

The examples reported show that, in natural convection, turbulence is directly produced by the positive buoyancy term (G_k) only in the presence of an imposed unstable thermal stratification, as occurs in the Rayleigh-Bénard problem or in internally heated cavities cooled from above). In all other cases, the production of turbulence is mediated by the shear term (P_k) and thus is only indirectly caused by natural convection; in some cases, a stable thermal stratification is established which counteracts the effects of shear and keeps turbulence levels moderate.

6.5   Turbulence in the Atmospheric Boundary Layer (ABL)

An important example of turbulent mixed convection occurs in the Atmospheric Boundary Layer (ABL), in which the wind-ground interaction always produces turbulence while the thermal stratification either produces or destroys it depending on whether it is unstable or stable (Nieuwstadt and van Dop 1982). In this section, we will derive a simple expression for the Richardson number in the ABL in order to assess the relative importance of wind and thermal gradients and classify the possible stability conditions.

In order to account for the effects of pressure on the vertical density stratification, the absolute temperature T must be replaced by the *potential temperature*

$$\theta = T\left(\frac{p_0}{p}\right)^{\frac{\gamma-1}{\gamma}} \tag{6.19}$$

in which $\gamma = c_p/c_v$ and p_0 is a reference pressure (e.g. 1 bar). The quantity θ is defined as the temperature that would be attained by a dry air mass at given absolute temperature T and pressure p if it expanded adiabatically from p to p_0. It is also called *adiabatic temperature* and denoted by T_{adiab}.

Based on this definition, on the equation of state of perfect gases ($p/\rho = R'T$) and on the equation of adiabatic transformations ($p/\rho^\gamma = $ constant), simple passages lead to

$$\frac{\partial \theta}{\partial z} \approx \frac{\partial T}{\partial z} + \Gamma_a \tag{6.20}$$

in which $-\Gamma_a = g/c_p \approx 0.01$ K/m is called *dry adiabatic temperature gradient*. The thermal stratification is unstable if $\partial \theta / \partial z < 0$, i.e. $\partial T / \partial z < -\Gamma_a$ (super-adiabatic temperature profile), and is otherwise stable (sub-adiabatic temperature profile). Further corrections can be introduced to account for air humidity.

Second, observe that in the ABL, denoting by x the wind direction and by z the vertical direction, the only significant velocity is that along x (u) while the only significant gradients are those along z. Starting from Eq. (6.17) for the gradient Richardson number, substituting θ for T and observing that, treating air as a perfect gas, $\beta_T = 1/T \approx 1/\theta$, one obtains

$$Ri_g = \frac{g}{\langle \theta \rangle \sigma_t} \frac{\frac{\partial \langle \theta \rangle}{\partial z}}{\left(\frac{\partial \langle u \rangle}{\partial z} \right)^2} \tag{6.21}$$

Let's now make the following assumptions:

$$\mu_t \frac{\partial \langle u \rangle}{\partial z} = \rho u_\tau^2 \tag{6.22}$$

(which links the vertical velocity gradient to the friction velocity u_τ via the turbulent viscosity μ_t);

$$-c_p \Gamma_t \frac{\partial \langle \theta \rangle}{\partial z} = q_w \tag{6.23}$$

(which links the vertical temperature gradient to the ground-to-air heat flux q_w via the turbulent thermal diffusivity Γ_t);

$$\mu_t = \kappa z \rho u_\tau \tag{6.24}$$

(which expresses the turbulent viscosity μ_t by a Prandtl mixing length model in which the distance z from the ground is chosen as the length scale, the friction velocity u_τ is chosen as the velocity scale and $\kappa \approx 0.42$ is von Karman's constant).

Based on Eqs. (6.22)–(6.24), and remembering that the turbulent Prandtl number is $\sigma_t = \mu_t / \Gamma_t$, Eq. (6.21) can be written

$$Ri_g = \frac{z}{L_{MO}} \tag{6.25}$$

in which the quantity

$$L_{MO} = -\frac{\rho c_p \langle \theta \rangle u_\tau^3}{\kappa g q_w} \qquad (6.26)$$

is called *Monin–Obukhov length*.

The Monin–Obukhov similarity theory aims at defining universal profiles of the turbulence quantities in the ABL, functions only of z/L_{MO} and of the temperature gradient.

Since $z/L_{MO} = Ri_g = -G_k/P_k$, in the case of Fig. 6.7 (stable stratification, $\partial T/\partial z > -\Gamma_a$ or $\partial \theta/\partial z > 0$) q_w is negative (i.e., the ABL transfers heat to the ground), L_{MO} is positive, Ri_g is everywhere positive and G_k is everywhere negative. Therefore:

- for $z/L_{MO} < 1$, $-G_k < P_k$ (G_k is negative and less than P_k in absolute value), so that production of k by shear exceeds destruction by stable stratification, and high turbulence levels are attained;
- for $z/L_{MO} > 1$, $-G_k > P_k$ (G_k is negative and larger than P_k in absolute value), so that suppression of k by stable stratification exceeds production by shear, and low turbulence levels, or even flow laminarization, are attained.

Typical values of L_{MO} are of the order of 10^2 m.

Contrariwise, in the case illustrated in Fig. 6.8 (unstable stratification, $\partial T/\partial z < -\Gamma_a$ or $\partial \theta/\partial z < 0$) q_w is positive (i.e. the ground transfers heat to the ABL), L_{MO} is negative, Ri_g is everywhere negative and G_k is everywhere positive. As a consequence, both G_k and P_k concur in the production of turbulence at all vertical locations, so that the ABL is strongly turbulent everywhere.

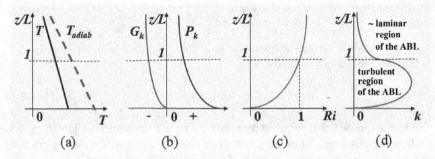

Fig. 6.7 Vertical profiles of different quantities in an ABL with stable thermal stratification ($L_{MO} > 0$). **a** temperature T and potential, or adiabatic, temperature T_{adiab}; **b** P_k (production of k by *shear*) and G_k (production of k by *buoyancy*); **c** Richardson number Ri; **d** turbulent kinetic energy k

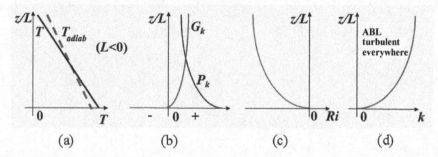

Fig. 6.8 Vertical profiles of different quantities in an ABL with unstable thermal stratification ($L_{MO} < 0$). **a** temperature T and potential, or adiabatic, temperature T_{adiab}; **b** P_k (production of k by *shear*) and G_k (production of k by *buoyancy*); **c** Richardson number Ri; **d** turbulent kinetic energy k

References

Abe K, Kondoh T, Nagano Y (1994) A new turbulence model for predicting fluid flow and heat transfer in separating and reattaching flows—I: flow field calculations. Int J Heat Mass Transf 37:139–151

Ampofo F, Karayiannis TG (2003) Experimental benchmark data for turbulent natural convection in an air filled square cavity. Int J Heat Mass Transf 46:3551–3572

Barhaghi DG, Davidson L, Karlsson R (2006) Large-eddy simulation of natural convection boundary layer on a vertical cylinder. Int J Heat Fluid Flow 27:811–820

Burns AD, Jones IL, Kightley JR, Wilkes NS (1989) Harwell-FLOW3D release 2 user manual. UKAEA report AERE-R (draft), July

Daly BJ, Harlow FH (1970) Transport equations in turbulence. Phys Fluids 13:2634–2649

Gatski TB, Bonnet J-P (2013) Compressibility, turbulence and high speed flow, 2nd edn. Academic Press, New York

Iida O, Kasagi N, Nagano Y (2002) Direct numerical simulation of turbulent channel flow under stable density stratification. Int J Heat Mass Transf 45:1693–1703

Lam CKG, Bremhorst KA (1981) Modified form of the $k - \varepsilon$ model for predicting wall turbulence. ASME J Fluids Eng 103:456–460

Nieuwstadt FTM, van Dop H (eds) (1982) Atmospheric turbulence and air pollution modelling. D Reidel Publishing Co, Dordrecht

Nobile E, Piller M, Stalio E (2000) Direct numerical simulation of turbulent mixed convection in internal flows. In: Proceedings of the 18th UIT heat transfer conference, Cernobbio, Italy, 28–30 June 2000

Nourgaliev RR, Dinh TN (1997) The investigation of turbulence characteristics in an internally-heated unstably-stratified fluid layer. Nucl Eng and Des 178:235–258

Rodi W (1987) Examples of calculation methods for flow and mixing in stratified fluids. J Geophys Res 92(C5):5305–5328

Tian YS, Karayiannis TG (2000a) Low turbulence natural convection in an air filled square cavity, Part I: thermal and fluid flow fields. Int J Heat Mass Transf 43:849–866

Tian YS, Karayiannis TG (2000b) Low turbulence natural convection in an air filled square cavity, Part II: the turbulence quantities. Int J Heat Mass Transf 43:867–884

Van Maele K, Mercy B (2006) Application of two buoyancy-modified $k - \varepsilon$ turbulence models to different types of buoyant plumes. Fire Safety J 41:122–138

Yan Z, Holmstedt G (1999) A two-equation turbulence model and its application to a buoyant diffusion flame. Int J Heat Mass Transf 42:1305–1315

Chapter 7
Transient Turbulence

Feynmann's algorithm for problem solving: (1) write down the problem; (2) think really hard; (3) write down the solution.
Murray Gell-Mann

Abstract In *transient*, or unsteady, turbulence the flow is not stationary, even in a statistical sense. The problems arising from applying Reynolds averaging to these situations are discussed, and more appropriate *phase* or *ensemble* averages are defined. A specific example of turbulence with longitudinal oscillations (reciprocating turbulent flow in a plane channel) is treated in detail, including also the theory of oscillating *laminar* flows. The influence of a temporal longitudinal acceleration on *transition* to turbulence in ducts, and the effect of *transverse* oscillations on drag, are also discussed.

Keywords Unsteady turbulence · Phase average · Ensemble average · Acceleration · Longitudinal oscillation · Transverse oscillation

7.1 Unsteadiness and Turbulence

In Chap. 5, we defined *Reynolds decomposition* on the basis of Eqs. (5.1)–(5.2). We also observed that, if the limit in Eq. (5.1) exists, it must be independent from the initial instant t, so that the mean field loses its time-dependence and $\langle \varphi(\mathbf{x}, t) \rangle = \langle \varphi(\mathbf{x}) \rangle$. Therefore, turbulence models based on Reynolds decomposition can strictly be applied only to statistically stationary turbulent flows (*steady* turbulence), while their application to *unsteady*, or *transient,* turbulence, in which the limit in Eq. (5.1) does not exist, is ill-founded.

Yet, several turbulent flows of great practical importance, both in Nature and in engineering, are *not* statistically stationary.

In a first class of such flows, unsteadiness is the result of some time-dependent (in particular, periodic) forcing. An obvious instance is the reciprocating flow in an internal combustion engine. Further examples are the turbulent flow in stirred vessels (Brucato et al. 1998; Alcamo et al. 2005) or around the blades of a turbine (Chima et al. 1993; Lardeau and Leschziner 2005), in which periodic oscillations (different in nature and frequency from turbulent fluctuations proper) are generated by the rotation of the stirrer or of the turbine disc. In Nature, several geophysical phenomena, including atmospheric and oceanic circulation, offer examples of approximately periodic turbulent flows, forced by the season cycle or, on a shorter time scale, by the day-night alternance (Deardorff 1974, 1980; Goulart et al. 2004).

When unsteadiness is forced and strictly periodic with period t_{per}, Reynolds averaging, Eq. (5.1), can be replaced by *phase averaging* (Hussain e Reynolds 1970):

$$\langle \varphi \rangle (\mathbf{x}, t) = \lim_{N \to \infty} \frac{1}{N} \sum_{k=0}^{N-1} \varphi \left(\mathbf{x}, t + k t_{per} \right) \tag{7.1}$$

In this case, the mean value $\langle \varphi \rangle$ and higher order statistics maintain their dependence upon time and are periodic with period t_{per} so that it is sufficient to study them in any interval $(0, t_{per})$.

In a second class of flows, statistical unsteadiness arises as an intrinsic feature of the solution, even in the presence of constant forcing and constant boundary conditions. An example is the turbulent flow past a backward-facing step (Aubrun et al. 2000): for some values of the Reynolds number, the shear layer separating the recirculation bubble from the free-stream flow flaps back and forth, so that the reattachment length oscillates about its time-mean value. Similar turbulent flows with spontaneous time dependence occur also when a fluid flows around obstacles, yielding, for sufficiently high Reynolds numbers, a turbulent vortex street (Murakami and Mochida 1995). Further examples of turbulent flows with time-dependent mean are encountered in natural convection (Di Piazza and Ciofalo 2000).

Clearly, in such transient turbulent flows, lacking an exact underlying periodicity, phase averaging does not apply. In certain cases, the time scales t_{trans} of the transient are well separated from the turbulence time scales, which are typically represented, as anticipated in Chap. 3, by a LETOT (Large Eddy TurnOver Time) δ/u_τ, δ being a length scale of the domain of interest (e.g. the half-height of a plane channel, as in Sect. 3.3, or the radius of a pipe) and u_τ the friction velocity. An example is the turbulent flow in municipal water piping in which a LETOT is 0.1–1 s while the flow rate may vary with typical time constants of minutes to hours. In such cases, frequency spectra will exhibit a *spectral gap* between the frequencies $(t_{trans})^{-1}$ and $(\delta/u_\tau)^{-1}$. The Reynolds formalism can be retained provided the limit in Eq. (5.1) is replaced by the average over a time interval larger than the largest time constant of turbulence $(\sim \delta/u_\tau)$ but shorter than the time scale t_{trans} of the transient.

In other problems, however, the separation of time scales does not hold; one may even meet problems in which a turbulent flow is forced at a frequency *higher* than the intrinsic frequencies of turbulence.

In all cases in which turbulence unsteadiness is present, but neither phase-averaging nor finite-time averaging can be applied, the most general possible decomposition must be applied, i.e. that based on *ensemble* averaging:

$$\langle \varphi \rangle (\mathbf{x}, t) = \lim_{N \to \infty} \frac{1}{N} \sum_{k=1}^{N} \varphi^{(k)}(\mathbf{x}, t) \tag{7.2}$$

in which $\varphi^{(k)}$ is the value of φ in the k-th *realization* of a given flow and the summation is made over an *ensemble* (set) of N of realizations, with N tending to infinity. Conceptually, the realizations are identical physical systems subjected to the same boundary conditions and forcing terms, which, however, develop quite different turbulent flow fields as the result of minute differences, e.g. in the initial conditions. Phase averaging can be regarded as a particular case of ensemble averaging, in which the *ensemble* consists of all available periods.

Like phase averaging, also ensemble averaging leaves the mean value $\langle \varphi \rangle$ still a function of space and time, but, unlike phase averaging, it can be applied to all turbulent flows. If applied to statistically stationary flows, it will yield the same (time-independent) results as Reynolds averaging; if applied to flows with a regular periodic forcing, it will yield the same results as phase averaging.

Turbulence unsteadiness may occur in the most different varieties. In regard to its *temporal* features, the kind most studied, because it lends itself to the powerful phase averaging treatment and, in the long run, is indeed a kind of permanent flow, is that arising from periodic oscillations of some forcing term. Oscillatory flows, and, in particular, reciprocating flows with zero mean flow rate, will be discussed in Sects. 7.2 and 7.3. The more general issue of the influence of acceleration on turbulence and transition to turbulence will be discussed in Sect. 7.4.

In regard to the *spatial* features of turbulence unsteadiness, oscillations or other kinds of unsteadiness may occur in all directions and may be caused either by pressure changes or by the physical motion of solid walls.

Longitudinal oscillations (affecting the main velocity component) have been amply studied in laminar flow for their relevance in physiological problems (pulsatile flow of blood in vessels); extensions of these studies to the turbulent flow case are more recent and relatively rare. Sections 7.2–7.4 of this chapter are mainly devoted to longitudinal oscillations/accelerations.

Spanwise (lateral or circumferential) oscillations in turbulent flow have been studied, mainly by numerical simulation, because they promise to reduce friction by interfering with the near-wall coherent structures of turbulence. Section 7.5 will discuss these phenomena.

Finally, *wall-normal* oscillations have received little attention so far.

A rather exhaustive, but unfortunately not very recent, review of theoretical, computational and experimental results on oscillatory flows, including studies of turbulent flow, was presented by Gündoğdu and Çarpinlioğlu (1999). Most research on oscillatory turbulent flows has regarded geometrically simple configurations, such as plane channels and circular ducts.

7.2 Laminar Oscillatory Flow

As a necessary introduction to the more difficult case of *turbulent* oscillatory flow, this section summarizes the known theoretical results for *laminar* oscillatory flow.

The simplest oscillatory flow is that studied in the so called *Stokes second problem* (Stokes 1850): the motion of a viscous fluid adjacent to an indefinite plane wall which oscillates harmonically along x with velocity

$$u_w(t) = u_0 \sin(\omega t) \tag{7.3}$$

The analytical solution found by Stokes for the fluid's velocity is:

$$u(y,t) = u_0 \exp(-y/l_S) \sin(\omega t - y/l_S) \tag{7.4}$$

in which y is the normal distance from the wall and $l_S = (2\nu/\omega)^{1/2}$. Today l_S is called *Stokes length*, and the near-wall region of thickness l_S is called *Stokes layer*. Equation (7.4) shows that the perturbation caused by the wall's motion decreases exponentially with y and penetrates a few Stokes lengths into the fluid.

Only slightly more complex is the oscillatory flow in a duct. The problem has been investigated for its relevance in physiology (Womersly 1955) but finds important applications also in the field of industrial engineering (Mackley and Stonestreet 1995). Often the flow is classified as *pulsatile* when the flow rate oscillates about a non-zero mean value, *reciprocating* when this mean value is nil.

In the following, we will assume the duct to be a plane channel of half height δ as in Fig. 3.1, with the origin of the coordinate y in the channel's midplane, and will first consider the case of *reciprocating* flow.

In the laminar regime, the flow is purely parallel, $\mathbf{u} = (u, 0, 0)$, and is a function of only y and time. The only Navier–Stokes equation (along x) reduces to

$$\frac{\partial u}{\partial t} = \nu \frac{\partial^2 u}{\partial y^2} + \frac{1}{\rho} f_V(t) \tag{7.5}$$

in which f_V is a forcing term (dimensionally, force per unit volume) which can be interpreted either as a pressure gradient or as a body force, as will be discussed below. In any case, f_V is assumed to vary harmonically in time as

$$f_V(t) = f_0 \cos(\omega t) \tag{7.6}$$

The hydrodynamic problem consists of determining $u(y, t)$ for any choice of δ, f_0, ω and fluid's physical properties (density ρ, viscosity $\mu = \rho\nu$). The main parameter controlling the solution is the ratio between the momentum diffusion time scale (viscous scale), δ^2/ν, and the period of the forced oscillation, $t_{per} = 2\pi/\omega$. It is customary to use the *Womersley number* α_W, proportional to the square root of the above ratio:

$$\alpha_W = \delta\sqrt{\omega/\nu} \tag{7.7}$$

If $t_{per} \gg \delta^2/\nu$ (slow oscillations, small α_W) the Stokes layer has time to grow along the channel's walls as the harmonic forcing term increases, and the u profile follows closely a sequence of stationary Poiseuille profiles $u(y,t) = f_V/(2\mu) \times (\delta^2 - y^2)$. Contrariwise, if $t_{per} \ll \delta^2/\nu$ (rapid oscillations, large α_W), instantaneous u profiles are affected by inertial terms and differ largely from the parabolic, steady-state shape; they may exhibit local maxima out of the midplane and also change sign along y. Maxima of u are smaller than the steady-state value $f_V\delta^2/(2\mu)$ and are out of phase with respect to the forcing term; amplitude damping and phase lag increase with α.

The analytical solution for $u(y, t)$ (Landau and Lifshitz 1959; Loudon and Tordesillas 1998) is

$$u(y, t) = \frac{f_0}{\omega\rho\beta_W}\{ [\sinh\varphi_1(y)\sin\varphi_2(y) + \sinh\varphi_2(y)\sin\varphi_1(y)]\cos\omega t$$
$$+ [\beta_W - \cosh\varphi_1(y)\cos\varphi_2(y) - \cosh\varphi_2(y)\cos\varphi_1(y)]\sin\omega t\} \tag{7.8}$$

in which

$$\varphi_1(y) = \frac{\alpha_W}{\sqrt{2}}\left(1 + \frac{y}{\delta}\right) \tag{7.9}$$

$$\varphi_2(y) = \frac{\alpha_W}{\sqrt{2}}\left(1 - \frac{y}{\delta}\right) \tag{7.10}$$

$$\beta_W = \cosh\left(\sqrt{2}\alpha_W\right) + \cos\left(\sqrt{2}\alpha_W\right) \tag{7.11}$$

The problem admits two distinct physical interpretations. On one hand, f_V can be regarded as a pressure gradient $f_V = -dp/dx$ applied to a fluid confined between fixed walls; in this case, u is the absolute velocity in the laboratory reference frame. As an alternative, more consistent with Stokes' second problem, the channel's walls can be assumed to move along x according to the harmonic law

$$x_w = x_0\cos(\omega t) \tag{7.12}$$

which implies velocity and acceleration

$$\dot{x}_w = -\omega x_0\sin(\omega t) \tag{7.13}$$

$$\ddot{x}_w = -\omega^2 x_0\cos(\omega t) \tag{7.14}$$

In this second interpretation, the channel can be assumed to be open at both ends to a uniform-pressure environment, so that no pressure gradient exists and the *true* force f_V is nil. However, if the resulting flow is described with respect to an accelerated (*non-inertial*) reference frame integral with the walls, then a fictitious acceleration, opposite to that of the walls, must be added to the RHS of the momentum equation. This is obtained by setting $f_0 = \rho \omega^2 x_0$ in Eq. (7.6). In this case, u must be interpreted as the velocity of the fluid relative to the walls, while its absolute velocity (in the laboratory reference frame) will be

$$u_{abs}(y, t) = u(y, t) + (t) = u(y, t) - \omega x_0 \sin(\omega t) \tag{7.15}$$

Note that the amplitude u_0 of the walls' velocity $\dot{x}_w$ is $\omega x_0 = f_0/(\rho \omega)$.

As an example of the results obtained for different values of the Womersley number, Figs. 7.1 and 7.2 report the cross-stream profiles of u (graph a) and u_{abs}

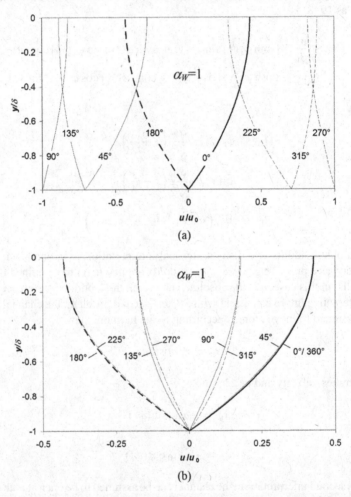

(a)

(b)

Fig. 7.1 Case $\alpha_W = 1$: cross stream velocity profiles, normalized by $u_0 = f_0/(\rho \omega)$, at different phase angles ωt. **a** Velocity relative to oscillating walls, equal to the velocity in a still channel under the effect of an oscillating pressure gradient; **b** absolute velocity between oscillating walls

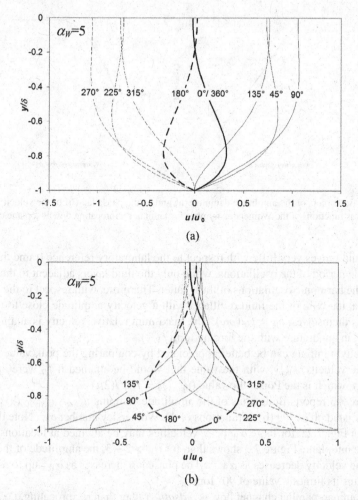

Fig. 7.2 Case $\alpha_W = 5$: cross stream velocity profiles, normalized by $u_0 = f_0/(\rho\omega)$, at different phase angles ωt. **a** velocity relative to oscillating walls, equal to the velocity in a still channel under the effect of an oscillating pressure gradient; **b** absolute velocity between oscillating walls

(graph b) at different phase angles ωt for $\alpha_W = 1$ and $\alpha_W = 5$, respectively. In both cases, velocities are normalized with respect to u_0 and their profiles are reported only for $y \leq 0$ since they are symmetric with respect to the midplane.

For low α_W, Fig. 7.1, the velocity perturbation caused by the oscillating walls (or by the oscillating pressure gradient) has time to diffuse to the bulk of the fluid during a semi-period, and the flow field can be described as a sequence of steady parabolic (Poiseuille) profiles in instantaneous equilibrium with the applied forcing. Velocity maxima are attained for $\omega t \approx 0$, in phase with the forcing term.

Results are quite different for high α_W, Fig. 7.2. A noteworthy feature of the solution is evidenced by the absolute velocity profiles in Fig. 7.2b: the central region

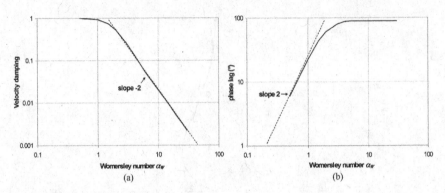

Fig. 7.3 Bode plots of the amplitude damping (**a**) and of the phase lag (**b**) of the velocity at the midplane as functions of the Womersley number for laminar reciprocating flow in a plane channel

of the fluid moves very little with respect to the laboratory reference frame through the whole period of the oscillations, while only the fluid layers adjacent to the walls follow the harmonic oscillations of these latter. Therefore, with respect to the walls, Fig. 7.2a, the bulk of the fluid oscillates with a velocity amplitude close to that of the walls themselves, $u_0 = f_0/(\rho\omega)$. The maximum relative velocity is attained for $\omega t \approx 90°$, in quadrature with the forcing term $f_V(t)$.

Velocity damping can be better appreciated by comparing the peak value of the midplane velocity, u_{peak}, with the value that would be attained if f_0 were applied statically, which is the Poiseuille value $(u_{peak})_0 = f_0\delta^2/(2\mu)$.

Figure 7.3 reports Bode plots of the amplitude damping $u_{peak}/(u_{peak})_0$ (a) and of the associated phase lag (b) as functions of the Womersley number α_W. Note that, as shown in Fig. 7.2a, for large α_W peak velocities may be attained at locations away from the midplane. Figure 7.3 shows that, for $\alpha_W > 2 \sim 3$, the amplitude of the peak midplane velocity decreases as α_W^{-2}. The phase lag increases as α_W^2 up to $\alpha_W \approx 1$ and attains its limiting value of 90° for $\alpha_W \approx 3$.

If a laminar, parallel channel flow is *pulsatile* rather than reciprocating, i.e., if the mean value of the flow rate is *not* nil, then the linear nature of the governing Eq. (7.5) allows the overall instantaneous flow to be obtained as the simple superposition of its constant component and its oscillatory component, the former being the Poiseuille solution and the latter being identical to the above solution for reciprocating flow (principle of superposition of effects). Therefore, the case of pulsatile flow does *not* present any real novelty with respect to reciprocating flow.

7.3 Turbulent Oscillatory Flow

Experimental results on turbulent oscillatory flows are not many, and all date back to some years ago. A vast experimental campaign was conducted between 1976 and 1995 at the University of Osaka in Japan by Ohmi and co-workers (Akao et al. 1986).

The authors investigated the flow of air in square or flat rectangular channels and in circular pipes induced by the reciprocating harmonic motion of a piston, thus realizing conditions of sinusoidally varying flow rate with zero mean. Averages and fluctuations were obtained as phase averages.

Curiously, similar but independent studies were conducted in the same years and a few hundred kilometres away at the Tokyo Institute of Technology by Hino and co-workers (Hino et al. 1983). The authors found that turbulence was triggered by a shear instability near the end of the acceleration phase, but maintained a low intensity up to the beginning of the deceleration phase. Velocity profiles followed the universal logarithmic law typical of turbulent flow in the deceleration phase but not in the acceleration phase, during which they resembled more laminar profiles. Contrariwise, velocity time spectra complied with the $-5/3$ slope typical of turbulent flow only in the acceleration phase, while they became steeper during deceleration.

Regime transition in reciprocating flows was discussed, among others, by Akhavan et al. (1991). Based on previous literature and on their own experimental results for circular ducts, they identified four flow regimes:

1. *purely laminar flow*, which strictly follows, in a plane channel, the theoretical behaviour described by Eqs. (7.8)–(7.11);
2. *disturbed laminar flow*, in which, especially in the acceleration phase, small velocity perturbations appear;
3. *intermittente turbulent flow*, in which turbulent *bursts* occur in the deceleration phase, while the flow reverts to ~laminar conditions in the acceleration phase;
4. *fully turbulent flow*, in which turbulence persists during the whole cycle.

According to the authors, the main parameter controlling flow regime transitions is the Reynolds number based on the Stokes length, $\mathrm{Re}_S = \bar{u}_{peak} l_S / v$, in which the overbar denotes average over the cross section while the subscript "peak" denotes the maximum value with respect to time.

This picture was fully confirmed by direct numerical simulations conducted by the author (Di Liberto and Ciofalo 2009). The computational domain was a rectangular "box" like that shown in Fig. 3.1, whose dimensions were 10δ along x and 4δ along z. No slip conditions were imposed at the top and bottom walls and periodicity conditions at the opposite side walls orthogonal to x and z. The computational grid was relatively coarse, including only $200 \times 80 \times 80$ ($x \times y \times z$) hexahedral finite volumes.

The equations solved were the usual continuity and Navier–Stokes ones written for a Newtonian, constant-property fluid subjected to a body force along x varying harmonically in time according to Eq. (7.6).

The amplitude f_0 of the forcing term, the wall shear stress $\tau_0 = \delta f_0$ and the friction velocity $u_{\tau,0} = (\tau_0/\rho)^{1/2}$ were use to build "static" wall units with respect to which mean and fluctuating velocity, as well as higher-order statistics (such as the Reynolds stresses and the turbulent kinetic energy k) were normalized. The friction velocity Reynolds number was defined as $\mathrm{Re}_{\tau,0} = u_{\tau,0}\delta/v$ and coincided with the channel's half-height in wall units, δ^+.

Besides *phase* statistics, well suited to this turbulent flow with periodic forcing, also *spatial* statistics were built by computing instantaneous averages and variances as functions of y and t over *homogeneous xz* planes (see remarks in Sect. 2.2). For selected test cases, both first- and second-order spatial and phase statistics were shown to be essentially identical, consistent with the remarks concerning Fig. 2.1. This allowed spatial statistics, simpler to compute, to be used *in lieu* of the more cumbersome phase statistics, which require a large number of periods to converge.

Figure 7.4 reports the velocity u^+ at three monitoring points during a whole oscillation period for four cases characterized by the same value of the "static" friction velocity Reynolds number $Re_{\tau,0} = 1964$ but by decreasing values of the Womersley number α_W (30, 20 15 and 10, respectively). The three monitoring points are located at dimensionless distances from one wall of 1964 (midplane, point 1), ~25 (production region, point 2) and ~350 (logarithmic region, point 3).

The four cases correspond to the four flow regimes classified by Akhavan et al. (1991). In particular, type III is characterized by the abrupt appearance of turbulent bursts near the end of the acceleration phase, followed by turbulence decay through the whole deceleration phase; turbulence is almost absent in the early stages of the acceleration phase that follows flow reversal. Therefore, turbulence is markedly asymmetric in time with respect to velocity peaks (which roughly correspond to the zeroes of the forcing term).

Type IV differs from type III in that turbulent bursts occur earlier and the quasi-laminar phase is reduced to a short interval following velocity reversal, but is

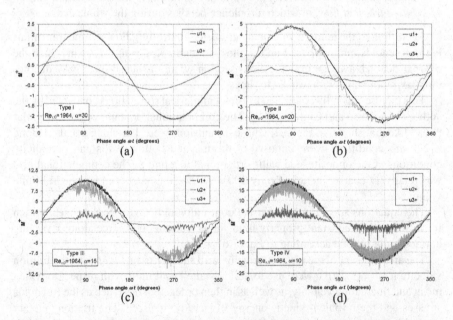

Fig. 7.4 Flow regimes in reciprocating channel flow predicted for $Re_{\tau,0} = 1964$ and different values of the Womersley number α_W. **a** Purely laminar; **b** disturbed laminar; **c** intermittent turbulent; **d** fully turbulent

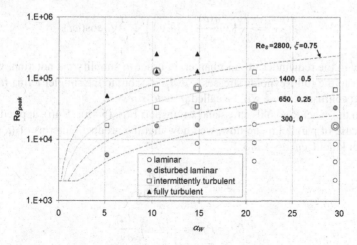

Fig. 7.5 Flow regime map in the (α_W, Re_{peak}) plane. See text for explanations

otherwise similar to type III in its general features, suggesting that a proper transition between the two types does not exist. Bursts are absent in type II regimes, characterized by low turbulence rather uniformly distributed over the whole cycle.

A similar sequence of flow types is encountered if the Womersley number α_W is kept fixed and the "static" friction velocity Reynolds number $\text{Re}_{\tau,0}$ is made to vary.

Figure 7.5 reports a flow regime map in the plane of α_W and Re_{peak}, which is the Reynolds number based on the hydraulic diameter (4δ) and the peak value (with respect to time) of the cross-section averaged velocity. The cases simulated are indicated with a different symbol according to the corresponding flow type; symbols enclosed in a circle denote the four test cases shown in the previous Fig. 7.4.

Isolines of the Stokes length Reynolds number Re_S are also reported, and it can be observed that they mark, at least approximately, the boundaries between flow types, thus suggesting that these can be classified based on the single parameter Re_S. This is in agreement with the criterion proposed by Akhavan et al. (1991) and mentioned above. Moreover, each value of Re_S is roughly associated with a certain value of the fraction of time spent by the fluid in the turbulent state, ξ, also reported in Fig. 7.5.

Thus, according to this map one has type I flow for $\text{Re}_S < 300$, corresponding to $\xi = 0$; type II flow for $300 < \text{Re}_S < 650$, corresponding to $\xi = 0$–0.25; type III flow for $650 < \text{Re}_S < 2800$, corresponding to $\xi = 0.25$–0.75; type IV flow for $\text{Re}_S > 2800$, corresponding to $\xi > 0.75$. Note that, for sufficiently high α (e.g. 30), the flow may remain laminar even at very high values of Re_{peak} (e.g. 10^4), essentially because the flow does not have time to develop turbulence structures between two consecutive flow reversals.

The behaviour illustrated by the time histories in Fig. 7.4 and the flow regime map in Fig. 7.5 can be interpreted in the light of the Reynolds equation governing turbulent parallel flow in a plane channel subjected to harmonic forcing:

$$\frac{\partial \langle u \rangle}{\partial t} = \frac{\partial}{\partial y}\left(\nu \frac{\partial \langle u \rangle}{\partial y} - \langle u'v' \rangle\right) + \frac{1}{\rho} f_0 \cos(\omega t) \qquad (7.23)$$

Note: in the remainder of this chapter, in order to simplify the notation, we will not refer to the full Reynolds stresses $-\rho\langle u'_i, u'_j \rangle$, but to the simpler terms $\langle u'_i, u'_j \rangle$. To avoid confusion, these will be called *pseudo-stresses*.

From the exact Reynolds stress and k transport Eqs. (5.45), (5.46), applied to this simple case of parallel mean flow, we know that the production terms of $\langle u'v' \rangle$ and turbulent kinetic energy k are:

$$P_{uv} = -\left\langle v'v' \right\rangle \frac{\partial \langle u \rangle}{\partial y} \qquad (7.24)$$

$$P_k = P_{uu} = -\left\langle u'v' \right\rangle \frac{\partial \langle u \rangle}{\partial y} \qquad (7.25)$$

respectively. Equation (7.25) shows that production directly affects only the longitudinal normal pseudo-stress $\langle u'u' \rangle$; turbulence energy is then transferred to the other normal components $\langle v'v' \rangle$ and $\langle w'w' \rangle$ by the redistribution terms (correlations between the fluctuations of pressure and of velocity gradients).

A simplified block diagram of the complex interactions between mean flow and turbulence is reported in Fig. 7.6, where only the most important terms are represented. The closed-loop nature of the interaction network is clear.

Under constant forcing, the closed-loop system in Fig. 7.6 would admit a steady-state (equilibrium) solution for all mean and turbulence quantities like $\langle u \rangle$ and $\langle u'v' \rangle$. Contrariwise, under the action of time-periodic forcing, the same quantities oscillate periodically. The various phase lags between mean flow and forcing term, or between

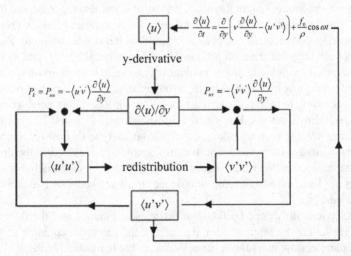

Fig. 7.6 Simplified block diagram of the closed-loop interactions between mean flow and turbulence in plane channel flow

fluctuations and mean flow, depend on the amplitude and frequency of forcing, as represented by $Re_{\tau,0}$ and α_W, respectively.

In the following, the relations between the various quantities in Fig. 7.6 will be illustrated in detail, on the basis of direct numerical simulations, for a type III (intermittent turbulent) flow characterized by $Re_{\tau,0} = 1389$, $\alpha_W = 15$. For this case, the behaviour of $u(t)$ is qualitatively similar to that reported in Fig. 7.4c for $Re_\tau = 1964$: fluctuations remain low during the acceleration phase, increase almost abruptly near the end of acceleration and remain significant during the whole deceleration phase, almost to vanish at flow reversal.

Figure 7.7 reports cross-stream profiles of various quantities (in "static" wall units) at phase intervals of $\pi/8$. For clarity purposes, profiles are shown only in the first half period; note that $\langle u \rangle (\omega t + \pi) = -\langle u \rangle (\omega t)$.

Graph (a) shows the mean velocity $\langle u \rangle$. As a general rule, $\langle u \rangle$ lags by almost $\pi/2$ behind the forcing term $f_V(t)$, as it would in a *laminar* reciprocating flow for the present, high value of the Womersley number α_W, see Fig. 7.3b. Profiles of $\langle u \rangle$ remain concave during the whole acceleration phase ($0 < \omega t < \pi/2$), turn flat at $\omega t = \pi/2$, when $\langle u \rangle$ attains its maximum value, and become convex during the deceleration phase ($\pi/2 < \omega t < \pi$). Slightly negative values of $\langle u \rangle$ are attained for $\omega t = \pi$. Velocity profiles at phases symmetrically placed with respect to $\pi/2$, e.g. $\omega t = \pi/8$ (acceleration) and $\omega t = 7\pi/8$ (deceleration), exhibit similar values in the midplane, but are totally different elsewhere. These results are agree with those reported by Hino et al. (1983).

Corresponding profiles of the turbulent kinetic energy k, graph (b), show that, for any ωt, k is significant only in the near-wall layers and very low in the central region of the channel. k is lowest shortly after the beginning of the cycle ($\omega t \approx \pi/8$) and remains low during the whole acceleration phase; peak values of k are attained in the early stages of the deceleration phase ($\omega t \approx 5\pi/8$), with a phase lag of $\sim\pi/8$ (20–25°) behind the peaks of $\langle u \rangle$, and at a distance from the walls close to the Stokes length $(2\nu/\omega)^{1/2}$. Both phase and location coincide with the findings of Scotti and Piomelli (2001). At further stages of the deceleration ($\omega t \approx 6\pi/8$ to π) near-wall values of k decrease rapidly but turbulence spreads into the central region.

Finally, graph (c) reports corresponding profiles of the turbulent shear pseudo-stress $\langle u'v' \rangle$. In regard to the amplitude of $\langle u'v' \rangle$, remarks similar to those made for k hold; the highest levels of $\langle u'v' \rangle$ are attained in the early stages of the deceleration phase (e.g. $\omega t = 5\pi/8$). In regard to the *sign* of $\langle u'v' \rangle$, it should be observed that, at instants symmetrically located with respect to $\pi/2$, e.g. $\omega t = \pi/8$ and $7\pi/8$, profiles of $\langle u \rangle$ are similar and forward-oriented, but profiles of the turbulent pseudo-stress $\langle u'v' \rangle$ are opposite.

For the same test case considered in Fig. 7.7 ($Re_{\tau,0} = 1389$, $\alpha_W = 15$ yielding type III, or intermittent turbulent, flow), Fig. 7.8 reports time histories of several mean and turbulence quantities at various distances y^+ from a wall, from ~6 to ~1389 (midplane). For symmetry reason, histories are reported only for half period and half channel.

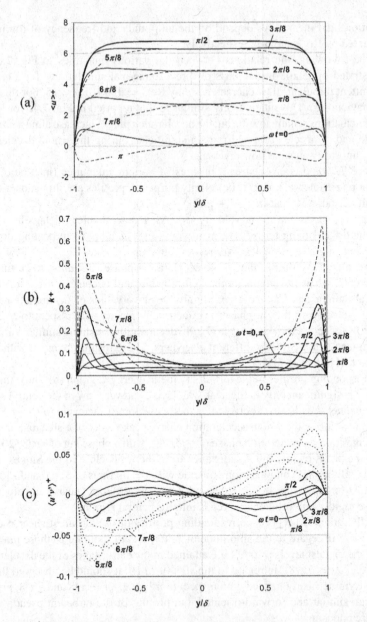

Fig. 7.7 Type III reciprocating channel flow ($Re_{\tau,0} = 1389$, $\alpha = 15$): cross-stream profiles in "static" wall units at different phase angles ωt. (a) mean velocity $\langle u \rangle$; (b) turbulent kinetic energy k; (c) turbulent shear pseudo-stress $\langle u'v' \rangle$

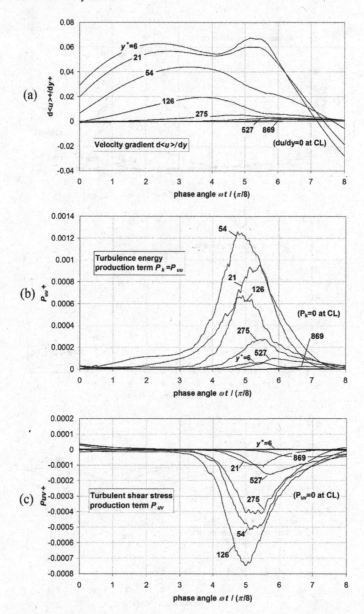

Fig. 7.8 Type III reciprocating channel flow ($\mathrm{Re}_{\tau,0} = 1389$, $\alpha_W = 15$): different quantities in "static" wall units during half cycle as functions of the phase angle at different distances y^+ from a wall. **a** Mean velocity gradient $\partial\langle u \rangle/\partial y$; **b** production term $P_{uu} = P_k$; **c** production term P_{uv}. (continues). **d** normal turbulent pseudo-stress $\langle u'u' \rangle$; **e** normal turbulent pseudo-stress $\langle v'v' \rangle$; **f** tangential turbulent pseudo-stress $\langle u'v' \rangle$

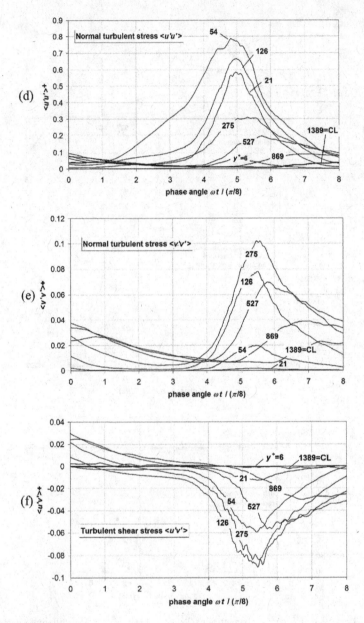

Fig. 7.8 (continued)

With the help of Fig. 7.8 and of the block diagram in Fig. 7.6, one can follow the complex phase relations which arise between different quantities and the corresponding production and destruction terms.

Consider a given location, say $y^+ = 54$. Given the closed-loop nature of the various inter-relations between the various quantities, one may start from any of them, for example the turbulent shear pseudo-stress $\langle u'v' \rangle$ in graph (f). This quantity exhibits a (negative) peak at $\omega t \approx 5.5 \times \pi/8$ while, for the same y^+, the mean velocity gradient $\partial \langle u \rangle / \partial y$ in graph (a) exhibits its peak at ~$3.5 \times \pi/8$. The product of these two quantities, which is the production term $P_{uu} = P_k$ of the normal turbulent pseudo-stress $\langle u'u' \rangle$ (and of the turbulent kinetic energy k), reported in graph (b), exhibits its peak value at a phase angle of ~$5 \times \pi/8$, intermediate between the two previous ones. The peak value of the normal turbulent pseudo-stress $\langle u'u' \rangle$, graph (d), is attained at $\omega t \approx 5 \times \pi/8$, almost exactly in phase with its production term P_{uu}, which indicates that the relevant production process does not imply any appreciable delay. Contrariwise, the other normal turbulent pseudo-stress $\langle v'v' \rangle$, graph (e), lags significantly behind P_{uu} and $\langle u'u' \rangle$, exhibiting its peak value at ~$5.5 \times \pi/8$; this is due to the time required for the re-distribution process. The product of $\langle v'v' \rangle$ (peaked at $5.5 \times \pi/8$) by $\partial \langle u \rangle / \partial y$ (peaked at ~$3.5 \times \pi/8$) is the production term P_{uv} of the turbulent shear stress $\langle u'v' \rangle$. It is reported in graph (c) and exhibits its (negative) peak at ~$5 \times \pi/8$, intermediate between the phase angles of its two factors. Finally, the turbulent shear pseudo-stress $\langle u'v' \rangle$, from which we started running the loop, lags slightly behind its production term P_{uv} and exhibits its own peak value at a phase angle of ~$5.5 \times \pi/8$.

In this complex interplay the mean velocity gradient $\partial \langle u \rangle / \partial y$ plays a pivotal role in rephasing the various turbulence components, compensating the delays implied by the production and redistribution processes.

Figure 7.8d makes it clear that turbulence production $P_{uu} = P_k$ is much more intense in the deceleration phase ($\omega t = \pi/2 - \pi$) than in the acceleration phase ($\omega t = 0 - \pi/2$). The reason for this can be found in the expression (7.25) for $P_{uu} = P_k$ and in the shape of the cross-stream profiles of mean velocity $\langle u \rangle$ and turbulent shear pseudo-stress $\langle u'v' \rangle$, reported in Fig. 7.7a, c, respectively.

In the acceleration phase, profiles of $\langle u \rangle$ in Fig. 7.7a do exhibit a large slope $\partial \langle u \rangle / \partial y$, but only very close to the walls, where corresponding values of $\langle u'v' \rangle$ in Fig. 7.7c are low; as a consequence, P_{uu} is small almost everywhere. In the deceleration phase, the slope $\partial \langle u \rangle / \partial y$ decreases slightly, but maxima of $\langle u'v' \rangle$ increase largely and, what is more important, spread over a broad region of the channel; as a consequence, local values of $P_{uu} = P_k$ increase and their integral across the channel increases even more, producing high turbulence levels.

In this scenario, temporal acceleration does not appear to play a direct role in turbulence balances; its effect is rather that of affecting the profiles of mean velocity, thus indirectly triggering a chain of consequences that translate into a modulation of turbulence intensity with the phase angle.

Under turbulent conditions, if the flow rate oscillates around a *nonzero* mean value (pulsatile flow), the principle of superposition of effects can no more be invoked, because the governing equations (complete Navier–Stokes) are not linear, even in the geometrically simplest configurations.

Thus, besides a "modulation" Reynolds number characterizing the amplitude of the oscillations and a Womersley number characterizing their frequency, a third parameter, e.g. a mean flow Reynolds number, is necessary to characterize the base flow (Stettler and Fazle Hussain 1986), and for each combination of these three parameters the time-dependent solution has to be computed as a whole. For any amplitude of the oscillatory forcing term f_0, turbulence is sustained by the base flow even at high frequencies, when a purely reciprocating flow, as discussed in the previous Sect. 7.3, would laminarize.

Scotti and Piomelli (2001) presented direct and large-eddy simulations of pulsatile flow in a plane channel. The authors found that, in a broad range of conditions, turbulence production attains a maximum at a phase angle ωt of ~$5\pi/8$ and at a distance from the walls of about one Stokes length, as defined in Sect. 7.2.

Shemer et al. (1985) conducted experiments on the turbulent pulsatile flow in circular ducts for $\alpha_W = 4.5$–15 and mean flow Reynolds numbers of 2900–7500. The ratio of oscillatory to mean flow rate did not exceed 35%. The authors found that turbulence intensity was larger in the deceleration phase than in the acceleration one, in which a tendency towards laminarization was observed. The phase lag of the mean velocity with respect to the forcing term was little affected by the Reynolds number but depended strongly upon the frequency of forcing (Womersley number α_W). Fluctuations, in their turn, were delayed with respect to the mean velocity; in the midplane this lag was ~$\pi/2$ for $\alpha_W = 12$ and decreased to a few degrees for $\alpha_W = 6.5$. Contrariwise, in the near-wall region fluctuations exhibited only a small phase lag with respect to the mean velocity.

7.4 Turbulence and Temporal Acceleration

The influence of *spatial* accelerations, i.e. velocity gradients, on turbulence is well known and can easily be explained by considering exact Reynolds stress balance equations such as Eq. (5.45). In particular, if a fluid flows along x in a strongly diverging duct, in which the most important Reynolds pseudo-stress is the normal one $\langle u'u' \rangle$ (positive) and the most important velocity gradient is $\partial \langle u \rangle / \partial x$ (negative), then the production term of $\langle u'u' \rangle$ in Eq. (5.45), P_{uu}, is approximately equal to $-2\langle u'u' \rangle \partial \langle u \rangle / \partial x$ (positive); normal fluctuations along x will grow by drawing energy from the mean flow (this turbulence energy will then be transferred also to the other directions by redistribution terms). Contrariwise, in a converging duct one will have $\partial \langle u \rangle / \partial x > 0$ and $P_{uu} < 0$, so that normal fluctuations (and the whole turbulence energy) will lose energy in favour of the mean flow. This mechanism is clearly discussed, for example, by Hinze (1975). Thus, a positive acceleration reduces turbulence while a negative acceleration (i.e., a deceleration) promotes turbulence.

The influence of *temporal* acceleration on turbulence is a different and more subtle issue, which (in the author's opinion) has not found in the literature a really satisfactory treatment so far. Experimental data on turbulent flows that are temporally accelerated or decelerated (Greenblatt and Moss 1999, 2004; He and Jackson 2000) suggest that the influence of the time derivative of the mean velocity is similar to that of the spatial axial derivative, i.e. temporally accelerating flows (in which the flow rate increases with time in absolute value, independent of the direction) exhibit low turbulence levels or even laminarization, whereas decelerated flows exhibit high turbulence levels.

Also the issue of *transition to turbulence* in accelerated flow has been the subject of experimental investigation for several decades (Kurokawa and Morikawa 1986; Lefebre and White 1991; Nakahata et al. 2007). The problem is also related to that of transient turbulence in oscillatory flows, as discussed, for example, by Di Liberto and Ciofalo (2011) and in Sect. 7.3 of this chapter. Although some aspects remain controversial, it is generally accepted that the main parameter controlling transition is the dimensionless, cross-section averaged, flow acceleration $\alpha = ad^3/\nu^2$ (in which d is the diameter of the duct, a is the acceleration $d\overline{u}/dt$ and ν is the fluid's kinematic viscosity), and that the transitional Reynolds number Re_{cr} increases approximately as $\alpha^{1/3}$.

Annus (2011) reviewed a large number of these studies; on the basis of literature results and of new experiments conducted at the large *Deltares* hydraulic test facility in Delft, he proposed a correlation which, once adapted to the present notation and clipped (somewhat arbitrarily) to 2100 in the limit of steady-state flow, can be written

$$\text{Re}_{cr} \approx \max\left(2100, \ 300\,\alpha^{1/3}\right) \tag{7.16}$$

Equation (7.12) implies that, for given fluid properties and (dimensioned) acceleration, Re^* is proportional to the pipe diameter. The influence of acceleration on transition is strong: for example, for water flowing in a 2 cm diameter pipe, even a moderate acceleration of 0.01 m/s^2 yields $\alpha \approx 10^5$ and $\text{Re}_{cr} \approx 14{,}000 \gg 2100$.

Most of the studies behind Eq. (7.16) were conducted under constant acceleration conditions. However, it seems reasonable to extend them also to constant pressure gradient conditions (decreasing acceleration), at least for moderate accelerations ($\alpha < 10^6$ in dimensionless form) for which velocity profiles at the instant in which $\text{Re} = \text{Re}^*$ do not significantly differ between the two conditions.

7.5 Turbulence and Spanwise Oscillations

The influence of spanwise (lateral or circumferential) oscillations on turbulence has mainly been studied in relation to the possibility that they reduce friction.

Pioneering experiments by Bradshaw and Pontikos (1985) and subsequent direct numerical simulations by Moin et al. (1990) showed that the abrupt imposition of a lateral pressure gradient on a turbulent boundary layer causes a reduction of turbulence intensity, followed by a recovery and a flow re-alignment in the near-wall region along the new oblique resulting direction. Based on these findings, Jung et al. (1992), by making the lateral pressure gradient a harmonic function of time and adjusting its frequency, obtained a reduction of up to 40% of longitudinal friction.

The same authors proved the equivalence between an oscillating pressure gradient and a harmonic oscillation of the wall with a spanwise velocity $w = w_0\sin(\omega t)$.

Similar results were obtained by Baron and Quadrio (1996). Quadrio and Sibilla (2000) later extended the previous computational results to the flow in circular ducts. They demonstrated by Direct Numerical Simulation that a circumferential oscillation of a pipe around its axis may reduce axial friction by up to 40%. They also showed that the oscillating flow in the near-wall region, once averaged in space over homogeneous cylindrical surfaces, strictly follows the analytical solution for oscillatory laminar flow in the Stokes layer.

Further studies (Laadhari et al. 1994; Choi and Graham 1998; Choi 2002; Luchini et al. 2006; Quadrio 2011) confirmed that wall-bounded flows, both in plane and circular geometry, exhibit a reduction of wall friction when they are subjected to lateral oscillations of the walls or to oscillatory lateral pressure gradients.

Most of the techniques that have been proposed to exploit this phenomenon are active (i.e., require an external mechanical energy source) and open-loop (i.e., without feedback). Their weak point is that the energy required to maintain the oscillation against the fluid's viscous resistance can be significant compared with the saving of pumping energy allowed by the reduction of axial friction.

For the same reason, the condition of maximum net energy saving does not coincide, in general, with that of maximum friction reduction. For example, Quadrio and Ricco (2004) carried out a parametric DNS study of turbulent channel flow with spanwise oscillating walls, letting the period t_{per}^+ and the maximum velocity w_0^+ of the harmonic oscillation vary independently. They found that friction reduction is largest for any given w_0^+ when t_{per}^+ is in the range 100–125, and increases monotonically with w_0^+ for any given t_{per}^+. However, the maximum net saving in pumping energy, equal to ~7%, is obtained in a limited region of the (t_{per}^+, w_0^+) plane, centred at $t_{per}^+ \approx 125$, $w_0^+ \approx 4.5$.

It is now widely accepted that lateral oscillations reduce axial friction by disturbing the wall turbulence cycle. This disturbance has been described either as the induction of a phase lag between low speed *streaks* and quasi-longitudinal vortices (Baron and Quadrio 1996), or as the creation of negative spanwise vorticity by interfering with the vortex stretching mechanism (Choi 2002). Experiments by Ricco (2004) have shown that the whole near-wall flow field is laterally stretched by the oscillation, which reduces the length of the *streaks* and enhances their spanwise spacing. Also Touber and Leschziner (2012) showed that lateral oscillations of a wall considerably deform the *streaks* and reduce the constribution of turbulence to wall shear stress. A DNS study by Ricco et al. (2012) has shown that wall oscillations directly affect the turbulence energy dissipation ε.

The beneficial effects of spanwise oscillations on friction decrease when the axial Reynolds number Re or the axial friction velocity Reynolds number Re_τ increase. For the same values of the oscillation amplitude and frequency studied by Quadrio and co-workers, DNS results by Choi et al. (2002) showed that the reduction of friction decreases by ~ 25% when Re_τ increases from 100 to 400. More recent DNS results (Ricco and Quadrio 2008; Touber and Leschziner 2012) confirmed a reduction of the effects of oscillation on friction (by 15% for Re_τ increasing from 200 to 400 and by 18% for Re_τ increasing from 200 to 500).

Wall oscillations need not to take the form of a rigid-body motion. For example, Viotti et al. (2009) simulated a steady-state forcing consisting of the lateral motion of the wall with velocity $w = w_0^+ \sin(2\pi x/\Lambda_x)$, constant in time but varying harmonically with the longitudinal coordinate x (stationary wave). This is, in a sense, the spatial counterpart of a temporal oscillation of the wall. These authors found that, for any given amplitude w_0^+, a wavelength Λ_x optimal for drag reduction exists, in analogy with the existence of an optimum value of the period $t_{per} = 2\pi/\omega$ in the case of an oscillating wall. The two optimum values of Λ_x and t_{per} turned out to be related by the near-wall axial advective velocity. Therefore, the basic mechanisms for drag reduction do not seem to change between temporal and spatial oscillations, although the net energy balance seems to be in favour of the latter (stationary wave).

Spanwise oscillations can also be created by oscillating body forces rather than by the motion of the walls. For example, Berger et al. (2000) performed DNS for the flow of an electrically conductive fluid at $Re_\tau = 100$ and showed that a lateral Lorenz force varying harmonically with time can reduce axial friction up to 40%.

Du and Karniadakis (2000) and Du et al. (2002) proposed a spatially non-uniform lateral forcing, consisting of spanwise-travelling waves of spanwise volume force confined to the viscous sublayer. For $Re_\tau = 150$, their DNS results showed a drag reduction of up to 30% and a significant alteration of the wall turbulence cycle, in which sinuous low-speed *streaks* disappeared almost completely, replaced by a large straight ribbon of low-speed fluid.

Zhao et al. (2004) replaced the spanwise-travelling waves of spanwise volume force of Du and co-workers by spanwise-travelling waves of spanwise wall velocity of the form $w = w_0 \sin(2\pi z/\Lambda_z - \omega t)$. Results were similar in terms of drag reduction and alterations of the flow statistics, but disappointing in terms of energy savings, which were negative for all the tested values of the parameters. Moreover, the prescribed wall velocity distribution would be quite hard to realize in real-world systems, requiring extensible walls.

Quadrio et al. (2009) and Quadrio and Ricco (2011) considered a wall velocity distribution of the form $w = w_0 \sin(2\pi x/\Lambda_x - \omega t)$, corresponding to longitudinally-travelling waves of spanwise velocity with a phase speed (wave celerity) $c = \omega\Lambda_x/2\pi$. The sign of c differentiates waves travelling forward and backward with respect to the mean fluid's motion. The functional form of $w(x, t)$ includes the limiting cases of rigidly oscillating walls (as in Quadrio and Sibilla (2000)) for $\Lambda_x \to \infty$ and of stationary waves, as in Viotti et al. (2009), for $\omega = 0$.

Based on a parametric DNS study, the authors obtained a map showing how drag varies with Λ_x and ω for given Re_τ and w_0^+. In particular, for $Re_\tau = 200$, $w_0^+ = 12$ and $\omega^+ > 0$ (forward-travelling waves), the largest drag reduction (~45%) is obtained along a steep line of approximate equation $\omega^+ = \Lambda_x^+$, while the line $\omega^+ = \Lambda_x^+$ roughly corresponds to the maximum drag increase (~20%). This region of maximum drag roughly corresponds to a celerity equal to the near-wall advective velocity, which shows that waves travelling forward with the same speed of the near-wall turbulence

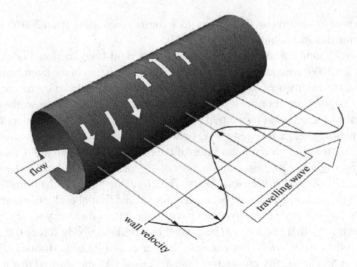

Fig. 7.9 Experimental realization of longitudinally-travelling waves of spanwise velocity. The required spatio-temporal distribution of wall speed is approximated by the independent rotation of adjacent sections of the duct. Reprinted from Auteri et al. (2010) with the permission of AIP Publishing

structures actually promote drag, whereas waves still travelling forward, but much more slowly, yield the largest drag reduction.

A wall velocity distribution of the kind studied computationally by Quadrio and co-workers (longitudinally-travelling waves of spanwise velocity) was actually experimentally realized by Auteri et al. (2010) in a circular duct. The required spatio-temporal modulation was obtained by dividing the wall into narrow sections and imposing each section a different rotational speed, suitably varying with time and with the axial location (Fig. 7.9).

On the whole, the effects of lateral oscillations on turbulent drag are sufficiently well understood. However, most DNS studies have been limited to moderate Reynolds numbers ($Re_\tau = 100$–200), while hypothetical industrial applications would probably involve much higher Reynolds numbers.

7.6   Conclusive Remarks on Transient Turbulence

Although the basic phenomena occurring in transient (and, in particular, oscillatory) turbulence are now well understood, our knowledge of this vast subject is still fragmentary. In particular:

- few results have been presented on *scalar transport* (including heat transfer) in unsteady and oscillatory turbulence;

- little is known of the effect of simultaneous oscillations in *more than one direction* (because of the strong nonlinearity of turbulence, the principle of superposition of effects cannot be invoked);
- wall oscillations *orthogonal* to their surface have received little attention so far; the literature offers only examples of more or less vaguely connected issues, such as compliant walls (Lee et al. 1993) or ultrasound forcing (Park and Sung 2005).

Yet, the field of the possible applications is very broad, and the potential benefits of a more complete and systematic knowledge of these phenomena are enormous. Besides the relatively well-explored issue of drag reduction, in which unsteadiness is explicitly imposed, suffice it to mention the correct prediction of friction and heat or mass transfer in pipes and devices installed on ground, naval or airborne vehicles, which are inevitably subjected to unwanted oscillations.

References

Akao F, Iguchi M, Ohmi M (1986) Velocity distribution in laminar phases for oscillatory rectangular duct flows with occurrence of transition. Bull JSME 29(254):2447–2454

Akhavan R, Kamm RD, Shapiro AH (1991) An Investigation of transition to turbulence in bounded oscillatory stokes flow. Part 1. Experiments; Part 2. Numerical simulations. J Fluid Mech 225:395–422 and 423–444

Alcamo R, Micale G, Grisafi F, Brucato A, Ciofalo M (2005) Large-Eddy simulation of turbulent flow in an unbaffled stirred tank driven by a Rushton turbine. Chem Eng Sci 60:2303–2316

Annus I (2011) Development of accelerating pipe flow starting from rest. Doctoral thesis. Civil Engineering, Tallinn University of Technology, TUT Press

Aubrun S, Kao PL, Boisson HC (2000) Experimental coherent structures extraction and numerical semi-deterministic modelling in the turbulent flow behind a backward-facing Step. Exp Thermal Fluid Sci 22:93–101

Auteri F, Baron A, Belan M, Campanardi G, Quadrio M (2010) Experimental assessment of drag reduction by traveling waves in a turbulent pipe flow. Phys Fluids 22:115103

Baron A, Quadrio M (1996) Turbulent drag reduction by spanwise wall oscillations. Appl Sci Res 55:311–326

Berger TW, Kim J, Lee C, Lim J (2000) Turbulent boundary layer control utilizing the Lorentz force. Phys Fluids 12(3):631–649

Bradshaw P, Pontikos NS (1985) Measurements in the turbulent boundary layer on an 'infinite' swept wing. J Fluid Mech 159:105–130

Brucato A, Ciofalo M, Grisafi F, Micale G (1998) Numerical prediction of flow fields in baffled stirred vessels: a comparison of alternative modelling approaches. Chem Eng Sci 53:3653–3684

Chima RV, Giel PW, Boyle RJ (1993) Algebraic turbulence model for three-dimensional viscous flow. In: Rodi W, Martelli F (eds) Engineering turbulence modelling and experiments 2. Elsevier, Amsterdam, pp 775–784

Choi K-S (2002) Near-wall structure of turbulent boundary layer with spanwise-wall oscillation. Phys Fluids 14:2530–2542

Choi K-S, Graham M (1998) Drag reduction of turbulent pipe flows by circular-wall oscillation. Phys Fluids 10:7–9

Choi J-I, Xu C-X, Sung HJ (2002) Drag reduction by spanwise wall oscillation in wall bounded turbulent flows. AIAA J 40:842–850

Deardorff JW (1974) Three-dimensional numerical study of the height and mean structure of a heated planetary boundary layer. Boundary Layer Meteorology 7:81–106

Deardorff JW (1980) Stratocumulus-capped mixed layers derived from a three-dimensional model. Boundary Layer Meteorology 18:495–527

Di Liberto M, Ciofalo M (2009) Numerical simulation of reciprocating turbulent flow in a plane channel. Phys Fluids 21:095106

Di Liberto M, Ciofalo M (2011) Unsteady turbulence in plane channel flow. Comput Fluids 49:258–275

Di Piazza I, Ciofalo M (2000) Low-Prandtl number natural convection in volumetrically heated rectangular enclosures—I. Slender cavity, AR = 4. Int J Heat Mass Transf 43:3027–3051

Du Y, Karniadakis GE (2000) Suppressing wall turbulence by means of a transverse traveling wave. Science 288:1230–1234

Du Y, Symeonidis V, Karniadakis GE (2002) Drag reduction in wall-bounded turbulence via a transverse travelling wave. J Fluid Mech 457:1–34

Goulart AG, Moreira DM, Carvalho JC, Tirabassi T (2004) Derivation of eddy diffusivities from an unsteady turbulence spectrum. Atmos Envir 38:6121–6124

Greenblatt D, Moss EA (1999) Pipe-flow relaminarization by temporal acceleration. Phys Fluids 11:3478–3481

Greenblatt D, Moss EA (2004) Rapid temporal acceleration of a turbulent pipe-flow. J Fluid Mech 514:66–75

Gündoğdu MY, Çarpinlioğlu MÖ (1999) Present state of art on pulsatile flow theory. Part 1: laminar and transitional flow regimes; Part 2: turbulent flow regimes. JSME Int J Ser B 42:384–397 and 398–410

He S, Jackson JD (2000) A study of turbulence under conditions of transient flow in a pipe. J Fluid Mech 408:1–38

Hino M, Kashiwayanagi M, Nakayama A, Hara T (1983) Experiments on the turbulence statistics and the structure of a reciprocating oscillatory flow. J Fluid Mech 131:363–400

Hinze JO (1975) Turbulence, 2nd edn. McGraw-Hill, New York

Hussain AKMF, Reynolds WC (1970) The mechanics of an organized wave in turbulent shear flow. J Fluid Mech 41:241–258

Jung WJ, Mangiavacchi N, Akhavan R (1992) Suppression of turbulence in wall-bounded flows by high-frequency spanwise oscillations. Phys Fluids A 4:1605–1607

Kurokawa J, Morikawa M (1986) Accelerated and decelerated flows in circular pipe (1st report, velocity profiles and friction coefficient). Bull JSME 29:758–765

Laadhari F, Skandaji L, Morel R (1994) Turbulence reduction in a boundary layer by a local spanwise oscillating surface. Phys Fluids 6:3218–3220

Landau LD, Lifshitz EM (1959) Fluid mechanics. Pergamon Press, Reading, MA

Lardeau S, Leschziner MA (2005) Unsteady RANS modelling of wake-blade interaction: computational requirements and limitations. Comput Fluids 34:3–21

Lee T, Fisher M, Schwarz WH (1993) Investigation of the stable interaction of a passive compliant surface with a turbulent boundary layer. J Fluid Mech 257:373–401

Lefebre PJ, White FM (1991) Further experiments on transition to turbulence in constant-acceleration pipe flow. ASME J Fluids Eng 113:428–432

Loudon C, Tordesillas A (1998) The use of the dimensionless Womersley number to characterize the unsteady nature of internal flow. J Theor Biol 191:63–78

Luchini P, Quadrio M, Zuccher S (2006) Phase-locked linear response of a turbulent channel flow. Phys Fluids 18:1–4

Mackley MR, Stonestreet P (1995) Heat transfer enhancement and energy dissipation for oscillatory flow in baffled tubes. Chem Eng Sci 50:2211–2224

Moin P, Shih TH, Driver D, Mansour NN (1990) Direct numerical simulation of a three-dimensional turbulent boundary layer. Phys Fluids A 2:1846–1853

Murakami S, Mochida A (1995) On turbulent vortex shedding flow past 2D square cylinder predicted by CFD. J Wind Eng Ind Aerodyn 54–55:191–211

Nakahata Y, Knisely CW, Nishihara K, Sasaki Y, Iguchi M (2007) Critical Reynolds number in constant-acceleration pipe flow. J Jap Soc for Exp Mech 7(2):142–147 (in Japanese)

Park YS, Sung HJ (2005) Influence of local ultrasonic forcing on a turbulent boundary layer. Exp Fluids 39:966–976

Quadrio M (2011) Drag reduction in turbulent boundary layers by in-plane wall motion. Phil Trans Roy Soc A 369:1428–1442

Quadrio M, Ricco P (2004) Critical assessment of turbulent drag reduction through spanwise wall oscillation. J Fluid Mech 521:251–271

Quadrio M, Ricco P (2011) The laminar generalized Stokes layer and turbulent drag reduction. J Fluid Mech 667:135–157

Quadrio M, Sibilla S (2000) Numerical simulation of turbulent flow in a pipe oscillating around its axis. J Fluid Mech 424:217–241

Quadrio M, Ricco P, Viotti C (2009) Streamwise-traveling waves of spanwise wall velocity for turbulent drag reduction. J Fluid Mech 627:161–178

Ricco P (2004) Modification of near-wall turbulence due to spanwise wall oscillations. J Turbul 5, article N24

Ricco P, Quadrio M (2008) Wall-oscillation conditions for drag reduction in turbulent channel flow. Int J Heat Fluid Flow 29:601–612

Ricco P, Ottonelli C, Hasegawa Y, Quadrio M (2012) Changes in turbulent dissipation in a channel flow with oscillating walls. J Fluid Mech 700:77–104

Scotti A, Piomelli U (2001) Numerical simulation of pulsating turbulent channel flow. Phys Fluids 13:1367–1384

Shemer L, Wygnanski I, Kit E (1985) Pulsating Flow in a Pipe. J Fluid Mech 153:313–337

Stettler JC, Fazle Hussain AKM (1986) On transition of the pulsatile pipe flow. J Fluid Mech 170:169–197

Stokes GG (1850) On the effect of the internal friction of fluids on the motion of a pendulum. Trans Cambridge Phil Soc IX:8–106. On the effect of the internal friction of fluids on the motion of a pendulum

Touber E, Leschziner MA (2012) Near-wall streak modification by spanwise oscillatory wall motion and drag-reduction mechanisms. J Fluid Mech 693:150–200

Viotti C, Quadrio M, Luchini P (2009) Streamwise oscillation of spanwise velocity at the wall of a channel for turbulent drag reduction. Phys Fluids 21:115109

Womersley JR (1955) Method for the calculation of velocity, rate of flow and viscous drag in arteries when the pressure gradient is known. J Physiol 127:553–563

Zhao H, Wu J-Z, Luo J-S (2004) Turbulent drag reduction by traveling wave of flexible wall. Fluid Dyn Res 34:175–198

Chapter 8
Transition to Turbulence

Life is pleasant. Death is peaceful. It's the transition that's troublesome
Isaac Asimov

Abstract This chapter is dedicated to *transition* to turbulence. Following a brief introduction on bifurcations, the scenario leading from steady laminar to chaotic (turbulent) conditions is discussed in some detail for specific configurations (toroidal pipe, serpentine pipe, spacer-filled channel), and intermediate steady, periodic and quasi-periodic regimes are identified.

Keywords Pitchfork bifurcation · Hopf bifurcation · Periodic flow · Quasi-periodic flow · Symmetry breaking · Travelling wave

8.1 Attractors, Transitions and Bifurcations

A forced dissipative dynamical system such as a flowing viscous fluid (or, more correctly, its mathematical representation), starting from an arbitrary initial condition P_0 in its phase space S of dimensionality N, will eventually get arbitrarily close to a subset A of this space, of dimensionality $M < N$ (not necessarily small!) and thus necessarily of zero measure in S, called the *attractor* of the dynamical system. This was briefly discussed in Sect. 2.7.

Each attractor is essentially characterized by its topological nature (fixed point, limit cycle, 2-torus, "strange attractor") and corresponds to a flow regime (stationary, periodic, quasi-periodic, chaotic, respectively) as sketched in Fig. 2.9.

A transformation of an attractor into another (of the same or of a different topological type) may occur when some control parameter (e.g. a Reynolds number Re) exceeds a critical value, and corresponds to a flow regime transition. Classifying the possible transitions is the object of *bifurcation* theory (Kuznetsov 2004).

One of the simplest bifurcations is the *pitchfork*, symmetry-breaking bifurcation occurring when a steady symmetric solution ceases to be stable while two steady solutions, asymmetric and mirror reflections of each other about some plane, become

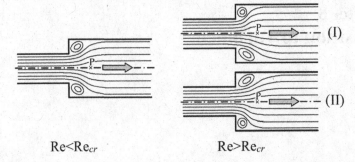

Re<Re$_{cr}$ Re>Re$_{cr}$

Fig. 8.1 Flow past a double symmetric flat duct expansion, illustrating a pitchfork (symmetry breaking) bifurcation

stable. A classic example, illustrated in Fig. 8.1, is the symmetry breaking occurring in the flow past a double symmetric flat duct expansion when the Reynolds number exceeds a certain critical value Re$_{cr}$ (Sobey and Drazin 1986).

For Re > Re$_{cr}$ one of the two solutions (I) and (II) will occur, "chosen" by the system practically at random on the basis of tiny irregularities present in the symmetric, low Re, solution.

A *bifurcation diagram* can be drawn by plotting an arbitrary quantity against the control parameter, for example the vertical velocity v at point P in Fig. 8.1 against the Reynolds number. Clearly, in the symmetric, low-Re, solution v is nil, while it is negative in solution (I) and positive in solution (II).

The bifurcation diagram may exhibit two different shapes, as sketched in Fig. 8.2a, b.

Case (a) is called a *supercritical* bifurcation: all stable solution branches lie in the half-plane where the control parameter is larger than its critical value, and the fixed-point attractor associated with the symmetric steady-state solution for Re < Re$_{cr}$ splits smoothly into a couple of fixed points, associated with the mirror-reflected

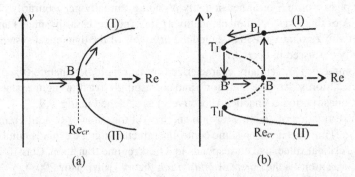

Fig. 8.2 Possible bifurcation diagrams for the pitchfork bifurcation. **a** Supercritical; **b** subcritical, exhibiting histeresys. Solid lines: stable branches; broken lines: unstable branches. T$_I$, T$_{II}$: turning points

asymmetric solutions (I) and (II) (only one of which will actually occur). In a real system, as the control parameter increases no abrupt transition is observed, but the solution, on leaving the bifurcation point B, evolves along one of the two alternative stable branches.

Case (b) is called a *subcritical* bifurcation: stable solutions exist for values of the control parameter both smaller and larger than its critical value. If the bifurcation point B is approached from the left (Re < Re_{cr}), on reaching B the solution leaves the symmetric branch, which becomes unstable for Re > Re_{cr}, and jumps abruptly onto either of the two stable asymmetric branches (for example, to the solution represented by P_I). In a physical system, an abrupt flow transition from the symmetric state to one of the two asymmetric steady states (I) or (II) is observed (compatibly with the system's inertia). However, if the system starts from a solution, say (P_I), lying on one of the two asymmetric stable branches and the control parameter (Re) is made to *decrease*, then the system remains on this branch even for Re < Re_{cr}, until the relevant turning point T_I is reached (corresponding to some Re < Re_{cr}), and then jumps abruptly back onto the symmetric branch (point B'). Thus, subcritical bifurcations may exhibit a closed-loop *histeresys* cycle (path B' → B → P_I → T_I → B'' in Fig. 8.2b).

.The *Hopf* bifurcation is associated with the system's attractor turning from a fixed point (steady state solution) to a limit cycle (periodic solution). In a physical system it corresponds to the appearance of tiny periodic time fluctuations in a previously stationary solution when a control parameter exceeds some critical value; the amplitude of the fluctuations grows as the control parameter increases further. The corresponding bifurcation diagram, Fig. 8.3, is better drawn in the three-dimensional space of two flow variables x, y and a control parameter λ.

A well known example of Hopf bifurcation is the appearance of a *von Karman vortex street* in the wake of an obstacle when the flow Reynolds number exceeds some critical value Re_{cr}.

More complex bifurcations mark the transition to quasi-periodic or chaotic states. Note that the transition from steady parallel-flow to turbulent flow in a straight duct occurs via a sequence of subcritical bifurcations so that intermediate (e.g. periodic or quasi-periodic) permanent regimes are never observed and can only exist transiently.

In the following, examples will be presented of flows crossing all or several possible intermediate states (steady, periodic, quasi-periodic and turbulent) as a

Fig. 8.3 Hopf bifurcation

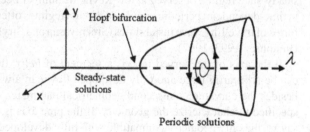

control parameter varies. In particular, the results of direct numerical simulations will be illustrated for three configurations:

- toroidal pipes;
- serpentine pipes;
- spacer-filled plane channels.

8.2   Transition in Toroidal Pipes

Consider a curved duct with circular cross section. Let $R = d/2$ and c be the pipe radius and the curvature radius, respectively, and let $\delta = R/c$ (curvature). The Reynolds number can be defined, as usual, as Ud/ν, in which U is the time and cross section-averaged velocity and ν is the fluid's kinematic viscosity.

The flow in curved pipes at sufficiently high Reynolds numbers is characterized by the presence of a secondary circulation in the cross section, caused by a disequilibrium between centrifugal forces and pressure gradients. This secondary flow appears whenever the Dean number $\mathrm{De} = \mathrm{Re}\ \sqrt{\delta}$ exceeds ~11.6 (Isachenko et al. 1974) which, for $\delta = 0.3$, corresponds to $\mathrm{Re} = 21.18$. Thus, the secondary circulation is present also in very low-Re flows (steady-state conditions). The secondary flow consists of a pair of counter-rotating vortices, called Dean vortices, whose peripheral velocity is of the order of $U\sqrt{\delta}$ (Boussinesq 1868; Dean 1927).

It has been known for a long time that in curved pipes the transition to turbulence is greatly delayed compared with straight pipes. A correlation for the critical Reynolds number that approximates well most of the results presented in the literature is (Di Piazza and Ciofalo 2011):

$$\mathrm{Re}_{cr} = 2100 \cdot \left(1 + 15\delta^{0.57}\right) \tag{8.1}$$

which, for example, yields $\mathrm{Re}_{cr} = 10{,}578$ for $\delta = 0.1$ and $\mathrm{Re}_{cr} = 17{,}958$ for $\delta = 0.3$, both values well above the critical Reynolds number in straight pipes (~2100).

The above critical Reynolds number is that for the onset of fully turbulent flow and is based on the experimental behaviour of global quantities such as the friction coefficient or the Nusselt number (Ito 1959; Cioncolini and Santini 2006). However, there are indications that, before the establishment of developed turbulence, the steady-state regime observed at low Reynolds number becomes unstable and results in time-dependent, periodic or quasi-periodic, regimes often characterized by various forms of travelling wave instabilities (Sreenivasan and Strykowski 1983; Webster and Humphrey 1993, 1997).

If the axis of the curved pipe is a segment of *helix*, then the length of the pipe can be arbitrarily large and fully developed flow can always be attained. However, besides the curvature δ, a second geometrical parameter, e.g. the *torsion*, has to be specified to characterize the geometry. If the pipe axis is imposed to lie in a plane, say xy, the only geometry compatible with fully developed flow is a *torus* (Fig. 8.4).

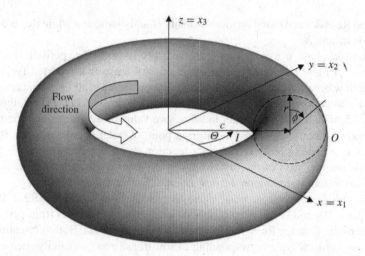

Fig. 8.4 Toroidal pipe

In a real-world system, the realizability of the flow in such a closed domain is indeed problematic, but means have been suggested in the literature to overcome this difficulty, as reviewed by Di Liberto and Ciofalo (2013). Of course, no realizability problem arises if numerical simulations are concerned.

For toroidal pipes, the author and co-workers (Di Piazza and Ciofalo 2011; Ciofalo et al. 2012) performed finite-volume direct numerical simulations which shed some light on the transition scenario, revealed the existence of intermediate, periodic and quasi-periodic, flow regimes regimes and suggested links with similar problems such as Taylor-Couette flow between concentric cylinders (Fenstermacher et al. 1979). Two values of δ were studied (0.1 and 0.3). For the sake of illustrating transition scenarios, it will be sufficient here to examine in detail the case $\delta = 0.3$.

Following a grid-independence study, a 3-D mesh including $\sim 3.34 \times 10^6$ finite volumes was chosen for this geometry (the corresponding surface mesh is shown in Fig. 8.4). The continuity and Navier-Stokes equations for a constant-density fluid (1.9), (1.35) were solved in time-dependent mode and the acceleration components a_i at the RHS of Eq. (1.35) were expressed so as to describe a driving pressure gradient $\mathbf{f}_V = \rho \mathbf{a}$ directed along the pipe axis. Therefore, the wall shear stress $\tau_w = R f_V$, the friction velocity $u_\tau = (\tau_w/\rho)^{1/2}$ and the friction velocity Reynolds number $\mathrm{Re}_\tau = u_\tau R/\nu$ were imposed, while the cross section-averaged mean velocity U and the corresponding bulk Reynolds number $\mathrm{Re} = Ud/\nu$ were obtained as part of the solution. In this study, Re_τ was made to vary in steps from 232 to 519, and the corresponding *bulk* Reynolds number varied from 4515 to 13,180.

Note that in any constant-section duct the Reynolds numbers Re and Re_τ and the Darcy friction coefficient f_D are related by

$$f_D = 32(\mathrm{Re}_\tau/\mathrm{Re})^2 \tag{8.2}$$

provided Re is defined based on mean velocity U and diameter d while Re_τ is defined based on u_τ and R.

A complete sequence of flow regimes, from steady-state (S) to periodic (P), quasi-periodic (QP) and chaotic (C), was obtained, similar to that reported for Taylor-Couette flow between concentric cylinders (Fenstermacher et al. 1979) and consistent with the classic transition sequence of Ruelle and Takens (1971). This is illustrated in Fig. 8.5, which reports the root mean square value of the oscillatory component of the axial velocity, u_s^{rms}, at a monitoring point of polar coordinates $r = 0.8R$, $\phi = -\pi/4$ in a generic cross section as a function of the departure from the critical Reynolds number Re_{cr} of the first instability. Symbols indicate computational test cases, and differ depending on the flow regime (S, P or QP).

A Hopf bifurcation from steady-state to periodic flow occurs for $Re = Re_{cr} = 4575$ (point H), and is followed by a secondary Hopf bifurcation from periodic to quasi-periodic flow for $Re = 5075 \approx Re_{cr} + 500$ (point H2). Both bifurcations are of the supercritical type, corresponding to soft transitions (a quantity such as u_s^{rms} changes continuously when Re crosses the relevant critical, i.e. bifurcation, value). As will be discussed below, the S $\rightarrow$ P transition is accompanied by a breaking of the instantaneous symmetry about the equatorial midplane of the torus; the resulting antisymmetry is preserved also in the subsequent QP regime. The transition to a chaotic (turbulent) regime is not shown in Fig. 8.5 and occurs at $Re \approx 8000 \approx Re_{cr} + 3500$.

The four flow regimes will be illustrated here by discussing a representative case for each regime. Table 8.1 summarizes the four test cases. The quantities reported

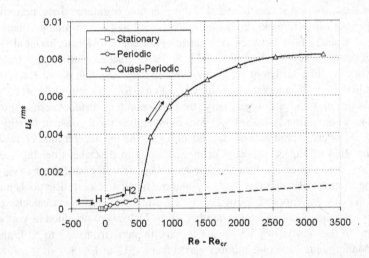

Fig. 8.5 Bifurcation diagram for a torus with $\delta = 0.3$. The abscissa is the difference between the Reynolds number Re and the critical Reynolds number Re_{cr} of the Hopf bifurcation from steady-state to time-dependent flow; the ordinate is the rms value of the oscillating axial velocity at a point of polar coordinates $r = 0.8R$, $\phi = -\pi/4$ in a generic cross section. H = Hopf bifurcation; H2 = secondary Hopf bifurcation. Solid lines: stable branches; broken lines: unstable branches

Table 8.1 Summary of four test cases representative of four flow regimes

Case	D3-S	D3-P	D3-QP	D3-C
Re_τ	234	247	290	361
Re	4556	4899	6128	8160
De	2495	2732	3356	4475
$f_D(\times 10^2)$	8.449	8.119	7.190	6.256

include the bulk Reynolds number (Re), the friction velocity Reynolds number (Re_τ), the Dean number (De) and the Darcy friction coefficient (f_D), Eq. (8.2).

In all the figures that follow, velocities are normalized by U (cross section and time average of the axial velocity), distances by R (minor radius of the toroidal pipe), time by R/U and frequency by U/R.

I—Steady-State Regime

For the steady-state case D3-S (Re = 4556), the flow field is illustrated in Fig. 8.6. Graph (a) reports a vector plot of the secondary flow, characterized by a couple of counter-rotating Dean vortices. Graph (b) reports the distribution of the axial velocity, characterized by a shift of the maximum towards the outer bend side.

The fluid flows towards the outer bend side along the equatorial midplane of the torus, returns towards the inner bend side along two symmetric boundary layers touching the upper and lower porions of the wall, and eventually forms the two counter-rotating Dean vortices which, for high Reynolds numbers like that considered

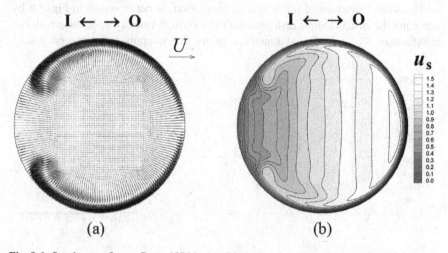

(a)   (b)

Fig. 8.6 Steady-state flow at Re = 4556 in an arbitrary cross section of a toroidal pipe with $\delta = 0.3$ (case D3-S). **a** Vector plot of the secondary flow (a vector of length U is reported); **b** contour plot of the axial velocity u_s, normalized by its average value U. Directions I (inner) and O (outer) are indicated

here, are strongly shifted towards the inner bend side while, at lower Re, they would be closer to the vertical midline.

Note that, under steady-state conditions, the flow field is identical in all cross sections, and the computational domain could consist of just a short tract of the pipe instead of the full torus.

II—Periodic Regime

As an example of periodic regime, consider the test case D3-P (Re = 4899).

Figure 8.7a reports time histories of $u_s - \langle u_s \rangle$ (fluctuating axial velocity) at two monitoring points, respectively located near the upper portion of the wall, within the upper secondary flow boundary layer, and near the centre of the upper Dean vortex in an arbitrary cross section. The time histories cover a dimensionless time interval of ~30. Periodicity is clear for both monitoring points; oscillations are larger, but still of small amplitude (~0.6% of the mean value) in the vortex region, and very small (~0.03% of the mean value) in the secondary flow boundary layer.

Figure 8.7b reports power spectra of the same velocities in the frequency domain, computed over a dimensionless time interval of ~140 (much larger than the interval in graph a). At both monitoring points the spectra exhibit a narrow peak at the dimensionless frequency $f^I \approx 0.238$; the harmonics $2f^I$ and $3f^I$ are well represented in the spectra in the Dean vortex region, but are negligible in the secondary flow boundary layer.

Because of the small amplitude of the oscillations, any visualization of the instantaneous flow field of the kind reported in Fig. 8.6 would *not* show any clear difference from a steady-state case.

The spatial structure of the flow field oscillations is better shown in Fig. 8.8 by reporting the instantaneous distribution of the vertical velocity u_z in the equatorial midplane of the torus. For symmetry, reasons, this quantity is nil in steady-state

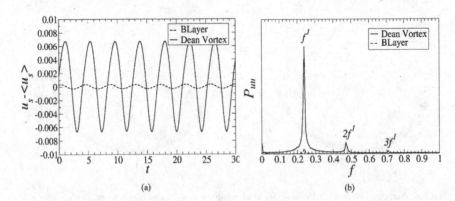

(a)   (b)

Fig. 8.7 Periodic case D3-P (Re = 4899): **a** time histories of the fluctuating axial velocity at two points, respectively located near the upper portion of the wall, within the upper secondary flow boundary layer, and near the centre of the upper Dean vortex; **b** corresponding spectra in the frequency domain (arbitrary units)

Fig. 8.8 Periodic case D3-P (Re = 4899): instantaneous distribution of the vertical velocity u_z (normalized by the mean velocity U) in the equatorial midplane of the torus. The direction and celerity of the travelling wave are indicated

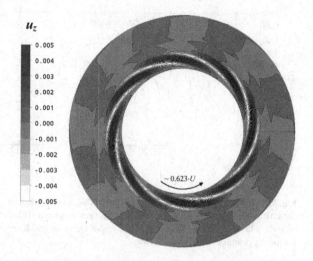

solutions. Contrariwise, in periodic flow the u_z distribution exhibits 8 repetitive cells, i.e. a wavenumber (number of waves included in the axis length $2\pi c$) $k^{\mathrm{I}} = 8$.

The whole distribution shown in the figure rotates rigidly around the vertical axis as a travelling wave, with a celerity (evaluated at the mean radius c) of ~0.623U. A perspective view—or, even better, a perspective animation of the flow—would show the travelling wave as a varicose modulation of both Dean vortex rings, with the modulations in the two rings shifted by half wavelength and thus in opposition of phase with each other (skew-symmetric, or anti-symmetric, flow). Since the wave celerity is lower than the mean fluid speed U, the wave travels *backward* with respect to the fluid (*trails* the fluid) in most of the torus' volume, with the exception of the near-wall region where the fluid speed is lower than 0.623 U.

All the periodic test cases simulated for $\delta = 0.3$ (including Reynolds numbers from 4605 to 5042) exhibit the same wavenumber $k' = 8$. Also the fundamental frequency f^{I} remains about constant (~0.238) through this whole interval.

III—Quasi-periodic Regime

This regime occurs when the Reynolds number exceeds ~5075 and is represented here by the test case D3-QP (Re = 6128).

Figure 8.9a reports time histories of the axial velocity fluctuation $u_s - \langle u_s \rangle$ at two monitoring points of a generic cross section, located, as in the previous case D3-P, one in the wall region (secondary flow boundary layer) and one near the centre of the upper Dean vortex. The time interval shown is ~30 in dimensionless form. Both fluctuations exhibit a non-periodic oscillatory behaviour, similar, at first sight, to a chaotic behaviour. However, power spectra of the same quantities in the frequency domain, computed over a dimensionless interval of ~220 and reported in Fig. 8.9b, contradict this impression. In fact, for both monitoring points the spectra exhibit narrow peaks at the two dimensionless frequencies $f^{\mathrm{I}} \approx 0.400$ and $f^{\mathrm{II}} \approx 0.165$, while secondary harmonics (multiple of the two above) are very small. Within the

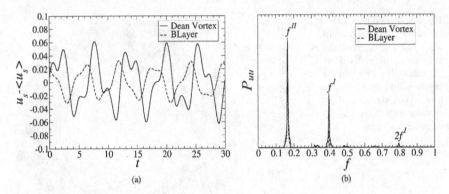

Fig. 8.9 Quasi-periodic case D3-QP (Re = 6128): **a** time histories of the axial velocity fluctuation at two monitoring points of a generic cross section, respectively located in the upper secondary flow boundary layer and near the centre of the upper Dean vortes; **b** corresponding power spectra in the frequency domain (arbitrary units)

frequency resolution of the present spectra, equal to the reciprocal of the histories' duration ($1/220 \approx 0.0045$ in dimensionless form), the two values 0.165 and 0.400 are incommensurate and thus characterize a quasi-periodic flow, in the sense discussed in Sect. 2.7. All other spectra computed for different quantities and at different locations exhibit only the two independent frequencies $f^{\mathrm{I}}, f^{\mathrm{II}}$ accompanied by their first few harmonics ($2f^{\mathrm{I}}, 3f^{\mathrm{I}} \ldots$ and $2f^{\mathrm{II}}, 3f^{\mathrm{II}} \ldots$).

Each of the two incommensurate frequencies $f^{\mathrm{I}}, f^{\mathrm{II}}$ (oscillation modes) is associated with a spatial structure that moves along the torus as a travelling wave. At most locations, the superposition of the two travelling structures gives rise to complex oscillations such as those shown by the time histories in Fig. 8.9a.

However, the region adjacent to the equatorial midplane happens to be interested almost exclusively by mode I and the near-wall region by mode II. Therefore, the spatial structure of mode I can be visualized by reporting, for example, the instantaneous distribution of the vertical velocity u_z on this equatorial plane, as in Fig. 8.10a. It is the same quantity reported, for the case D3-P of periodic flow, in Fig. 8.8; in the present case, however, the distribution is much different and the wavenumber is $k^{\mathrm{I}} = 7$ instead of 8. The celerity of the travelling wave is $\sim 1.19U$, slightly larger than the fluid's mean velocity; therefore, the disturbance wave moves forward with respect to the fluid (i.e., leads the fluid) in low-speed regions (e.g. near the walls), but moves backward with respect to the fluid (i.e., trails the fluid) in regions of high axial velocity (e.g. near the torus axis).

The fundamental frequency of mode I (~ 0.400) is much larger than the single fundamental frequency of the periodic case D3-P (~ 0.238); therefore, the transition from periodic to quasi periodic flow, occurring when the Reynolds number exceeds ~ 5000, is accompanied not only by the appearance of a second oscillation mode (mode II), but also by a strong increase of the fundamental frequency associated with the varicose modulation of the toroidal Dean vortices. This increase in frequency is mainly due to the strong increase of the linear celerity of the travelling wave (from

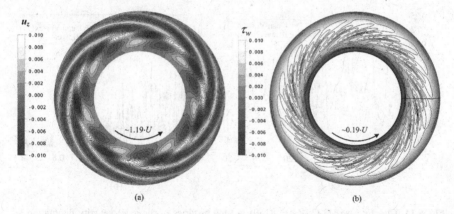

Fig. 8.10 Quasi-periodic case D3-QP (Re = 6128): **a** instantaneous distribution of the vertical velocity u_z in the equatorial midplane; **b** instantaneous distribution of the wall shear stress on the surface of the torus. Direction and celerity of travelling waves I and II are indicated

~$0.623U$ to ~$1.19U$), only partly compensated by the reduction from 8 to 7 of the wave number.

In the generic cross section, mode II manifests itself as a couple of vortex streets arising in the upper and lower boundary layers of the secondary flow. They move *against* the mean secondary flow, and thus from the inner (I) to the outer (O) bend side. As mode I, also mode II is instantaneously antisymmetric with respect to the equatorial midplane. The spatial structure of mode II is illustrated in Fig. 8.10b, which shows the instantaneous distribution of the wall shear stress on the surface of the duct, where the contribution of mode I is negligible. For the present case, this distribution includes $k^{II} = 18$ repetitive cells and rotates rigidly around the torus' axis with a linear celerity of ~$0.19U$, much less than the celerity of wave I and the mean velocity of the fluid.

Summarizing, in the quasi-periodic case D3-QP the time-dependent component of the flow field can be described as the superposition of two independent systems of travelling waves. Each consists of a spatially repetitive structure of k cells (modo) which rotates rigidly with its own celerity around the axis of the torus and in the same direction as the fluid's mean motion.

Mode I is the less energetic of the two modes and mainly affects the recirculating flow region. In the generic cross section, it takes the form of a weak periodic pulsation of the Dean vortices, while, in a three-dimensional perspective, it consists of a rigidly rotating spatial modulation of the ring vortices.

Mode II mainly interests the secondary flow boundary layers and is the more energetic of the two. In the generic cross section it appears as a couple of vortex streets which are generated at the edge of the Dean vortices and move from the inner to the outer bend side, opposite to the mean flow direction in the secondary flow boundary layers. The whole time-dependent component of the flow field is instantaneously anti-symmetric with respect to the equatorial midplane.

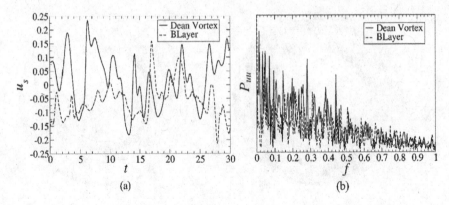

Fig. 8.11 Chaotic case D3-C (Re = 8160): **a** time histories of the axial velocity fluctuation at two monitoring points of a generic cross section, respectively located in the upper secondary flow boundary layer and near the centre of the upper Dean vortes; **b** corresponding power spectra in the frequency domain (arbitrary units)

As the Reynolds number varies in the quasi-periodic range from ~5075 to ~8000, the wave number k^I of mode I decreases from 8 to 7, while the corresponding dimensionless frequency f^I increases from 0.275 to 0.400. The wave number k^{II} of mode II increases from 10 to 18; the corresponding dimensionless frequency f^{II} increases from 0.035 to 0.165 up to Re = 6128 and then decreases for larger Re.

IV—Chaotic Regime

The highest Reynolds number for which a quasi-periodic solution was obtained was 7859 for the present curvature $\delta = 0.3$. A slight increase of Re above this value led to completely chaotic (turbulent) solutions. As an example of early chaotic flow, consider the test case D3-C at Re = 8160.

Figure 8.11a shows time histories of the fluctuating axial velocity $u_s - \langle u_s \rangle$ at the same monitoring points considered for P and QP flows in the previous Figs. 8.7a and 8.9a, and taken over a dimensionless time interval of 30.

Corresponding frequency spectra, computed from histories protracted for a dimensionless time of ~200, are reported in Fig. 8.11b. In both locations, velocity power spectra exhibit a continuous distribution of frequencies, typical of turbulent flow. Some peaks stand out against the background, but there is no trace of the modal frequencies $f^I \approx 0.4$, $f^{II} \approx 0.165$ observed in quasi-periodic cases at only slightly lower Reynolds number. Similar spectra are obtained if different quantities and locations are considered.

Also for this chaotic flow case, as for the periodic and quasi-periodic cases considered above, the time-averaged flow field in a generic cross section would not exhibit significant differences with respect to the steady-state case D3-S in Fig. 8.6. This shows that the transition to unsteady, even turbulent, regimes does not affect much the time averages.

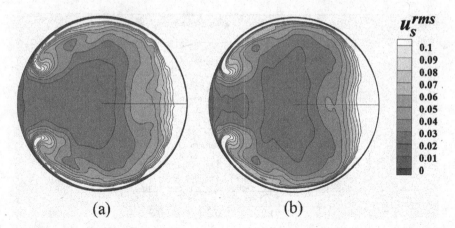

Fig. 8.12 Root-mean-square value of the axial velocity fluctuation (normalized by U) for chaotic flow at $\delta = 0.3$. **a** Re $= 8160$; **b** Re $= 13,180$

Figure 8.12a reports a contour map of the root-mean-square (rms) value of the axial velocity fluctuation, u_s^{rms}, for the present case D3-C. The most intense fluctuations are localized in the outer bend region. A further increase of Re to 13,180, Fig. 8.12b, does not significantly change the u_s^{rms} distribution, showing that the chaotic regime just above the transition value of ~8000 exhibits already the features of fully turbulent flow.

Also for the curvature $\delta = 0.1$ steady, periodic, quasi-periodic and chaotic regimes are obtained. However, the transition sequence is different and includes a subcritical bifurcation, with the associated histeresys cycle, from steady-state to time-periodic flow.

Based on the author's DNS predictions, literature results and asymptotic arguments, a tentative flow regime map in the (Re,δ) plane is sketched in Fig. 8.13.

The map accounts for the qualitative experimental findings of Sreenivasan and Strykowski (1983) at $\delta = 0.058$ and for the measurements of Webster e Humphrey (1997) at $\delta = 0.055$, but interprets the oscillatory regimes defined "periodic" by these authors as actual instances of quasi-periodic flow. The map accounts also for the fact that, for $\delta = 0$ (straight pipes), a direct transition from steady-state to turbulent flow occurs at Re ≈ 2100. Solid lines represent transitions occurring for increasing Re, while broken lines correspond to transitions occurring only for decreasing Re. Regions indicated as P-BW and QP-BW (for "BackWard") can be reached only by letting Re decrease starting from higher-Re solutions.

Schematic bifurcation diagrams corresponding to different intervals of the curvature δ are sketched in the lower part of the figure; in the proximity of $\delta = 0.3$ the diagram is qualitatively equal to that reported in Fig. 8.5. Details in the intermediate region between the curvatures 0.1 and 0.3 (shaded area in the figure) are purely hypothetical and based on a principle of maximum simplicity.

Figure 8.13 reports also the criterion for transition to turbulence in Eq. (8.1). For any curvature δ, it predicts a transitional Reynolds number much larger than that

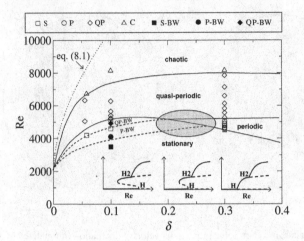

Fig. 8.13 Tentative flow regime map for toroidal pipes in the Re-δ plane. Symbols: computational results (S = steady-state, P = periodic, QP = quasi-periodic, C = chaotic, i.e. turbulent). S-BW, P-BW and QP-BW indicate S, P or QP solutions obtained only by letting Re decrease. Solid lines: transitions for increasing Re; broken lines: transitions for decreasing Re. Bifurcation diagrams corresponding to different intervals of δ are sketched. The shaded central region needs a deeper investigation

for transition to chaotic flow predicted in the present study and consistent with the literature. This shows that, as anticipated above, Eq. (8.1) and similar correlations do not identify the onset of chaotic flow, but rather the attainment of turbulence levels high enough to affect global quantities such as the friction coefficient. In geometrical configurations exhibiting important secondary flows, pressure drop can be much higher than in straight pipes even under conditions of steady ("laminar") flow; the onset of turbulence simply introduces a further dissipation term that may well give a minor contribution to the overall pressure drop.

8.3   Transition in Serpentine Ducts

The geometry considered in this Section, a repetitive unit of which is shown in Fig. 8.14, is a duct of circular cross section with radius $R = d/2$ made of consecutive U-bends of alternate curvature (serpentine pipe). Like the torus, it is characterized by a single shape parameter $\delta = R/c$ (curvature) but, unlike the torus, it is easily realizable and is actually encountered in engineering applications (Vashisth et al. 2008).

Ciofalo and Di Liberto (2017) presented DNS predictions for flow and heat transfer in the computational domain of Fig. 8.14, focussing on the loss of stability of the base (low-Re) flow and on the transition to time-dependent and eventually chaotic regimes. Four values of the curvature δ were examined (0.2, 0.3, 0.4 and 0.5); for

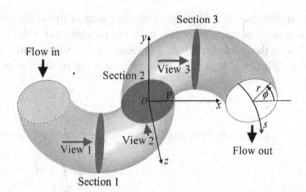

Fig. 8.14 Repetitive unit of a serpentine pipe, consisting of two U-bends with opposite curvature. The figure shows the cross sections 1–3 and the monitoring point P, where most results will be reported, and the corresponding view directions. Alternative coordinate systems (x, y, z) and (s, r, ϕ) are also indicated

each δ, the friction velocity Reynolds number $\mathrm{Re}_\tau = u_\tau R/\nu$ was made to vary in steps starting from low-Re solutions exhibiting stationarity and top-down symmetry. The Prandtl number was assumed to be unitary. A study of low-Re flow, including different Prandtl numbers and geometries, had been presented in a previous work (Ciofalo and Di Liberto 2015).

The hydrodynamic equations solved were, as for the torus discussed in the previous Section, the continuity and Navier–Stokes equations for a constant-property fluid (1.9), (1.34). No slip conditions were imposed at the walls. Fully developed flow was simulated by the "unit cell" approach, i.e. by imposing translational periodicity between inlet and outlet faces and expressing the acceleration components a_i at the RHS of Eqs. (1.34) so as to describe a driving pressure gradient $\mathbf{f}_V = \rho\mathbf{a}$ directed along the axis. Therefore, as for the torus, the wall shear stress $\tau_w = Rf_V$, the friction velocity $u_\tau = (\tau_w/\rho)^{1/2}$ and the friction velocity Reynolds number $\mathrm{Re}_\tau = u_\tau R/\nu$ were imposed, while the cross section-averaged mean velocity U and the corresponding bulk Reynolds number $\mathrm{Re} = Ud/\nu$ were obtained as part of the solution. As in all constant-section ducts, the Darcy friction coefficient f_D, Re and Re_τ are related by Eq. (8.2).

Heat transfer was governed by the energy equation for a constant-property fluid (1.39). A uniform heat flux q_w (entering the fluid) was imposed at the walls. In order to simulate thermally fully developed flow, inlet–outlet periodicity was imposed also to temperature and the volumetric power density q''' in Eq. (1.39) was written as $q''' = -(2\langle q_w \rangle/R) \times (u_s/U)$ (negative for a heated fluid), u_s being the local axial velocity component. This "unit cell" treatment is discussed in detail in the papers cited (Ciofalo and Di Liberto 2015, 2017). By adopting this approach, the computed p and T are the *periodic* components of pressure and temperature, to which large-scale uniform gradients along the axial direction **s** have to be added if the "true" p and T are required.

For a generic cross section of the duct, the bulk temperature T_b is defined as

$$T_b = \frac{1}{\dot{m}c_p} \int_A \rho c_p u_s T \, dA \qquad (8.3)$$

in which $A = \pi R^2$ is the (constant) area of the cross section and $\dot{m} = \rho U A$ is the mass flow rate. Note that, with the present "unit cell" treatment, T_b does not vary with the axial coordinate s in virtue of the compensative term q'''.

The *local* heat transfer coefficient h can be defined as

$$h = \frac{q_w}{T_w - T_b} \qquad (8.4)$$

The *average* heat transfer coefficient could be defined as

$$h_{avg}^{(1)} = \frac{1}{S} \int_S h \, dS \qquad (8.5)$$

in which S is the total area of the wall. However, in complex flows and under imposed heat flux conditions, the wall temperature may be locally equal, or even lower, than the bulk temperature T_b, so that h and $h_{avg}^{(1)}$ may become negative or exhibit singularities. A more robust definition of the average heat transfer coefficient, which does not suffer from this problem, is

$$h_{avg}^{(2)} = \frac{q_{w,avg}}{T_{w,avg} - T_b} \qquad (8.6)$$

in which $q_{w,avg}$ and $T_{w,avg}$ are the surface averages of wall heat flux and temperature.

The local heat transfer coefficient can be made dimensionless as a local Nusselt number $\mathrm{Nu} = h2R/\lambda$, and average Nusselt numbers $\mathrm{Nu}_{avg}^{(1)}$, $\mathrm{Nu}_{avg}^{(2)}$ can similarly be obtained from $h_{avg}^{(1)}$ and $h_{avg}^{(2)}$.

The numerical simulations were conducted using hexahedral grids with a number of finite volumes varying from ~1.8×10^6 for $\delta = 0.5$ to ~4.6×10^6 for $\delta = 0.2$. Results for steady-state flow were validated against experimental data of Wojtkowiak and Popiel (2000) for the Darcy friction coefficient.

The geometrical symmetries of the problem translate into some symmetry properties of the solution. These are better formulated using the "internal" coordinates s, r, ϕ and the associated velocity components u_s, u_r, u_ϕ instead of Cartesian coordinates x, y, z and components u, v, w.

Axial periodicity always implies for the generic quantity φ (velocity component, pressure or temperature) the translational symmetry

$$\text{SP:} \quad \varphi(s, r, \phi, t) = \varphi(s + 2\pi c, r, \phi, t) \qquad (8.7)$$

For sufficiently low Reynolds numbers, the solution (base flow) satisfies also the stationarity condition, or time-symmetry ST: $\partial\varphi/\partial t = 0$ and two further spatial symmetries:

$$\text{SU:} \quad \varphi(s, r, \phi) = \pm\varphi(s + \pi c, r, \pi - \phi) \qquad (8.8)$$

$$\text{SM:} \quad \varphi(s, r, \phi) = \pm\varphi(s, r, -\phi) \tag{8.9}$$

in which the minus sign applies when $\varphi = u_\theta$ and the plus sign for any other quantity. SU expresses the fact that, apart from translations and reflections, the solution is identical in the two U-bends that make up the computational domain. SM is the top-down symmetry with respect to the duct's equatorial midplane $y = 0$.

More complex symmetries may become important at higher Reynolds numbers.

In the following, for the sake of brevity, only results for the smallest and largest curvatures $\delta = 0.2$ and 0.5 will be discussed. Time will be made dimensionless as tU/d. The generic velocity component u_i will be made dimensionless as u_i/U, U being the time- and cross section-averaged axial velocity. Temperature T will be made dimensionless as $(T - T_b)/\Delta T_\lambda$, in which $\Delta T_\lambda = q_w d/\lambda$ is the conductive temperature scale.

Case $\delta = 0.2$ (low curvature)

I—Base Steady-State Flow

For the relatively low curvature $\delta = 0.2$, the base steady-state flow exhibiting symmetries SU and SM was obtained up to $Re_\tau \approx 29$. Figure 8.15a shows the iso-surface $u_s = U$ (axial velocity = mean velocity) for $Re_\tau = 25$. The pipe wall is shown as a semi-transparent surface.

In a constant-curvature pipe, such as the torus considered in Sect. 8.2, the velocity maximum would be shifted towards the outer bend side. In the present case, however, due to the periodic curvature change, a phase shift exists between the axial variation of

Fig. 8.15 Selected results for case $\delta = 0.2$, $Re_\tau = 25$ (symmetric steady-state flow at $Re \approx 206$). **a** Iso-surface in which the axial velocity equals its average value; **b** local Nusselt number distribution on the pipe wall

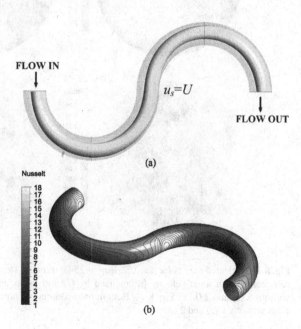

the geometry and that of the velocity distribution, so that velocity maxima approach the wall with a lag of more than $\pi/2$ behind the geometry. In no cross section the flow is under conditions of local equilibrium between pressure gradients and centrifugal forces as it would occur in a constant-curvature pipe.

Figure 8.15b shows the local Nusselt number distribution on the pipe wall for the same case. Regions of intense heat transfer occur where the $u_s = U$ isosurface comes closest to the wall, and thus are located in the entrance region of each U-bend on the inner bend side, while most of the outer bend side exhibits low values of Nu. This is in contrast with what is observed in constant-curvature curved pipes, e.g. tori or helical coils, and in single bends, where heat or mass transfer maxima are located on the outer bend side (Berger et al. 1983; Xin and Ebadian 1997; Janssen and Hoogendoorn 1978; Di Piazza and Ciofalo 2010).

Figure 8.16 reports selected results for the same case $\delta = 0.2$, $\mathrm{Re}_\tau = 25$. The graphs in the top row show contours of axial velocity u_s (normalized by the mean velocity U) superimposed on vector plots of secondary flow in cross sections 1 and 2 of Fig. 8.14. In section 1 the u_s maximum is still close to the axis, while in section 2 it has moved towards the outer bend side. The secondary circulation, with two characteristic Dean

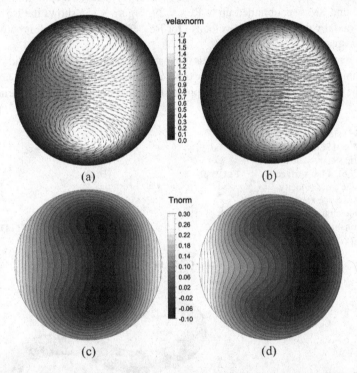

Fig. 8.16 Selected results for $\delta = 0.2$, $\mathrm{Re}_\tau = 25$ (symmetric steady-state flow at Re ≈ 206). Top row: contours of axial velocity (normalized by U) and vector plots of secondary flow in cross sections 1 (a) and 2 (b) of Fig. 8.14. Bottom row: contours of normalized temperature in the same cross sections 1 (c) and 2 (d)

vortices, is similar in shape and intensity in the two sections, with some shift of the recirculation centres towards the outer bend side from section 1 to section 2.

The graphs in the bottom row are contour plots of the dimensionless temperature in the same cross sections. Remarks similar to the above ones apply also for this quantity. At the present Re_τ symmetries SU and SM hold, so that both the velocity and the temperature distributions repeat themselves identically (apart from translations and reflections) in the two U-bends and are strictly top–bottom symmetric with respect to the equatorial midplane.

II—Steady Asymmetric Flow

For $\delta = 0.2$, letting Re_τ increase beyond ~ 29 led to a breaking of the spatial top–bottom symmetry SM and to the establishment of a new steady-state, top–bottom asymmetric, flow regime which still exhibits symmetry SU. The transition from steady symmetric to steady asymmetric flow was of the "hard" type, associated with a sub-critical pitchfork bifurcation: the flow was either symmetric or markedly asymmetric, and cases with just a small amount of asymmetry were not found.

Figure 8.17 reports the behaviour of the vertical velocity component v (normalized by U) at a monitoring point during the transient caused by the stepwise increase of the friction velocity Reynolds number Re_τ from 25 to 30 at $t = 0$. The monitoring point is that shown as point "P" in Fig. 8.11; it is located on the equatorial midplane in the outer region of section 2 at $s = 0$, $r = 0.85R$, $\phi = \pi$.

The plot shows that, following a transient oscillatory behaviour, v settles to a value of ~$0.16U$. Oscillations start almost immediately after the step increase of Re_τ, but their amplitude grows exponentially and remains negligible during the first 100–150 time units d/U. The transient is completed after ~$300d/U$.

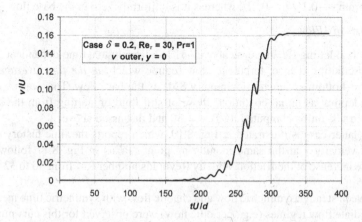

Fig. 8.17 Case $\delta = 0.2$: time history of the vertical velocity v at a monitoring point located in the equatorial plane following a stepwise increase of the friction velocity Reynolds number Re_τ from 25 to 30, showing symmetry breaking from symmetric to asymmetric steady-state flow (subcritical pitchfork bifurcation)

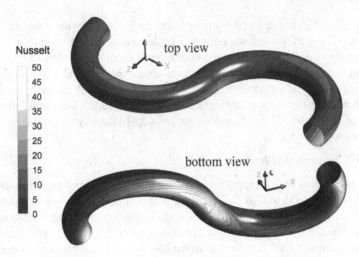

Fig. 8.18 Case $\delta = 0.2$, $\mathrm{Re}_\tau = 30$ (asymmetric steady-state flow at Re≈273): Nusselt number distribution on the pipe wall, as viewed both from top and from bottom to show loss of top–down (SM) symmetry. Note that symmetry SU is still retained

Figure 8.18 reports the distribution of the local Nusselt number Nu on the pipe wall for $\mathrm{Re}_\tau = 30$; two views, one from above and one from below, are reported to evidence the strong top–bottom asymmetry of the distribution. However, it can be observed that symmetry SU between the first and the second U-bend still applies.

Contours of axial velocity or temperature, as well as vector plots of the secondary flow in cross sections of the pipe, confirm the top–bottom asymmetry of the solution. Similarly, maps of the distribution of the vertical velocity component v on the equatorial midplane under the new steady-state conditions show that this quantity ranges from $\sim -0.1U$ to $\sim 0.2U$, whereas it is uniformly zero in the base flow.

III—Unsteady Flow

For $\delta = 0.2$, letting Re_τ increase above ~31 led directly to a time-dependent, irregularly oscillating (hence, turbulent) flow regime which, *in the time average*, was again top–bottom symmetric (symmetry SM) as the base flow. This was observed either adopting as initial conditions those of still fluid or starting from the steady asymmetric solution computed for $\mathrm{Re}_\tau = 30$ and discussed above.

This latter case is illustrated in Fig. 8.19, which reports the time history of the vertical velocity v at the same monitoring point "P" as in Fig. 8.17 following a stepwise increase of the friction velocity Reynolds number Re_τ from 30 to 32 at $t = 0$.

Between steady asymmetric flow and chaotic flow with symmetric time mean, no intermediate flow regimes (e.g., periodic flow) were observed for this curvature.

Some features of the chaotic flow computed for $\mathrm{Re}_\tau = 32$ are illustrated in Fig. 8.20, which reports the distribution of the local Nusselt number on the pipe wall at consecutive instants separated by a dimensionless interval $\Delta t U / d$ of ~ 1.43. The

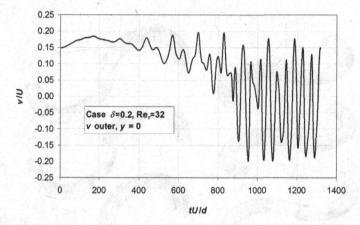

Fig. 8.19 Case $\delta = 0.2$: behaviour of the vertical velocity xxx at a monitoring point in the equatorial plane following a step increase of Re_τ from 30 to 32 at $t = 0$, showing transition from asymmetric steady-state flow to chaotic (turbulent) flow with symmetric time means

left column reports views from top, the right one from bottom. Transient maxima of Nu travel downstream and vanish after about one half bend length; circumferentially, they tend to occur in the proximity of the top and bottom generatrices of the curved pipe, rather than in the proximity of the equatorial midplane as observed for steady-state symmetric flow at $Re_\tau = 25$. For the *instantaneous* flow and temperature fields symmetries SU-SM are replaced, albeit only approximately, by the *anti*-symmetry condition

$$\text{SU}': \quad \varphi(s, r, \phi) = \varphi(s + \pi c, r, \phi - \pi) \qquad (8.10)$$

where φ is either a scalar quantity (p, T) or one of the velocity components u_s, u_r, u_ϕ.

Maps of axial velocity or temperature and vector plots of the secondary flow in a cross section at consecutive instants show that the flow field oscillates irregularly, with a pseudo-period of 5 time units d/U, and that the flow is instantaneously top–bottom asymmetric while it recovers symmetries SM and SU in the time averages.

The further increase of Re_τ does not change the qualitative behavior of the solution, but, as expected, causes larger and more irregular fluctuations. For example, Fig. 8.21 shows contours of the normalized axial velocity and vector plots of the secondary flow in cross Sect. 1 at consecutive instants for $Re_\tau = 50$ (yielding $Re \approx 570$). Frames are separated by a dimensionless time interval $\Delta t U/d$ of ~1.

For the same case ($\delta = 0.2$, $Re_\tau = 50$), power spectra of vertical velocity and temperature (not shown for brevity) exhibit a large number of energy-containing frequencies as is typical of turbulent flow, but follow poorly the $-5/3$ power law that would be characteristic of fully developed turbulence, which is not surprising taking account of the low Reynolds number (~570).

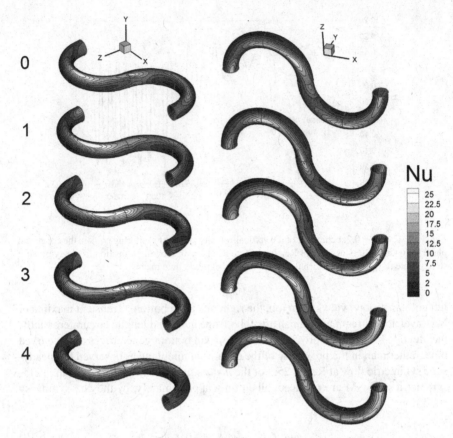

Fig. 8.20 Nusselt number distribution on the pipe wall at different instants for $\delta = 0.2$, $Re_\tau = 32$ (early chaotic flow at Re $\approx$ 300). The dimensionless time interval $\Delta t U/d$ between snapshots is ~1.43. Left: view from top; right: view from bottom. Note that (anti)symmetry SU′, Eq. (8.10), is approximately exhibited

Summarizing, the results for $\delta = 0.2$ indicate the following sequence of flow regimes and bifurcations (by "symmetric" we mean here top-down symmetry about the midplane): steady symmetric base flow → (subcritical pitchfork bifurcation) → steady asymmetric flow → chaotic flow with symmetric time mean, evolving smoothly to fully turbulent flow.

Case $\delta = 0.5$ (high curvature)

1—Base Steady-State Flow

Up to $Re_\tau \approx 28$, also the case $\delta = 0.5$ admitted steady-state solutions exhibiting symmetries SU and SM. Figure 8.22a shows the iso-surface $u_s = U$ (axial velocity = mean velocity) for $Re_\tau = 25$ (Re $\approx$ 142). Its shape is similar to that discussed for $\delta = 0.2$, with a similar phase shift between the spatially-periodic geometry variation

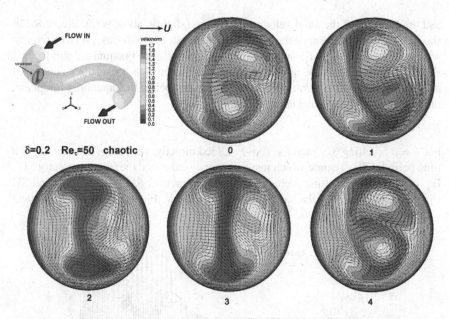

Fig. 8.21 Case $\delta = 0.2$, $Re_\tau = 50$ (chaotic flow at $Re \approx 570$): contours of axial velocity (normalized by U) and vector plots of secondary flow in cross section 1 (as indicated in the inset) at consecutive instants, separated by a dimensionless time interval $\Delta t U/d$ of ~1

Fig. 8.22 Selected results for case $\delta = 0.5$, $Re_\tau = 25$ (symmetric steady-state flow at $Re \approx 142$). **a** Iso-surface in which the axial velocity equals its average value; **b** Nusselt number distribution on the pipe wall

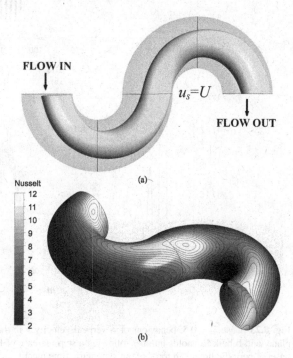

and the location of the axial velocity maximum (slightly above $\pi/2$). Figure 8.22b shows the Nu distribution on the wall for the same steady-state case. As a difference with the case $\delta = 0.2$ at the same $\text{Re}_\tau = 25$ (Fig. 8.15), Nu maxima are still located shortly downstream of the inflection cross section between consecutive bends, but are now peaked at two symmetric points above and below the equatorial midplane, rather than on the midplane itself.

II—Periodic Flow

For $\delta = 0.5$, letting Re_τ increase above ~28 led directly, via a Hopf bifurcation, to a time periodic flow regime which instantaneously exhibited the (anti)symmetry SU', Eq. (8.10), but in the time averages satisfied symmetries SM (top–bottom) and SU (first–second bend) like the base flow. As an example, Fig. 8.23 reports the vertical

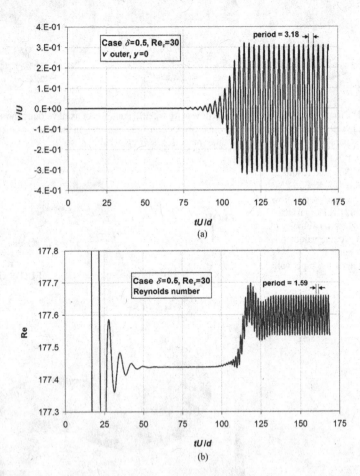

Fig. 8.23 Case $\delta = 0.5$: behaviour of **a** vertical velocity v in the outer region of the equatorial plane, and **b** bulk Reynolds number following a step increase of Re_τ from 25 to 30, showing the onset of periodic flow with top–bottom symmetric time mean

velocity v at the monitoring point "P" (a) and the bulk Reynolds number (b) as functions of time following a stepwise increase of the friction velocity Reynolds number Re_τ from 25 to 30 at $t = 0$.

Starting from steady state conditions at $Re \approx 142$, the stepwise increase of the forcing term causes large flow rate fluctuations, which last ~50 time units d/U and then settle to an apparent stationary equilibrium. However, this intermediate condition is soon destabilized by the growth of initially tiny oscillations, which eventually result in the breakdown of steady-state flow and the onset of periodic flow after ~130–140 time units. Once periodic flow is achieved, the period of the oscillations is ~1.59 time units for Re and twice as long (~3.18 time units) for the anti-symmetric local variable v. The relative amplitude of the oscillations is rather large (~$0.3U$) for the local velocity, but very small (less than 0.04%) for the Reynolds number.

The instantaneous antisymmetry SU' of the periodic flow is made evident by Fig. 8.24, which reports the distribution of the local Nusselt number on the pipe wall at a generic time, as viewed both from the top and from the bottom. Time averaging yields a symmetric Nu distribution which is similar to that reported in Fig. 8.22b for this curvature at $Re_\tau = 25$, and is not shown here for the sake of brevity.

Unlike in the previous case at lower curvature (0.2), the first bifurcation was now of the super-critical ("soft") type: arbitrarily small departures from the critical value of Re_τ led to correspondingly small amplitudes of the fluctuations.

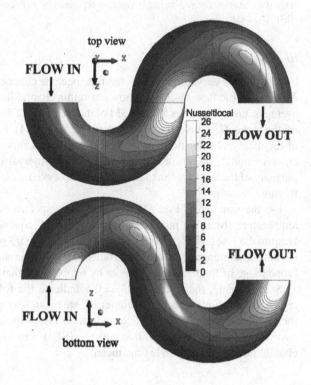

Fig. 8.24 Case $\delta = 0.5$, $Re_\tau = 30$ (periodic flow with symmetric time mean at $Re \approx 178$): instantaneous distribution of the local Nusselt number on the pipe wall at a generic time, as viewed from top and from bottom. Note (anti)symmetry SU', see Eq. (8.10)

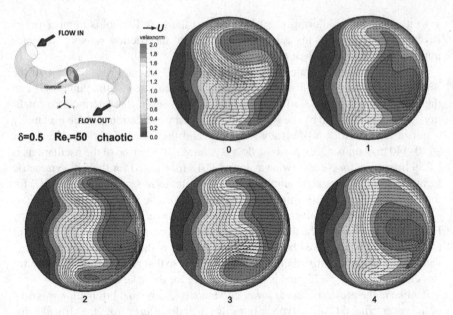

Fig. 8.25 Case $\delta = 0.5$, $Re_\tau = 50$ (chaotic flow with symmetric time mean at Re≈324): contours of axial velocity (normalized by U) and vector plots of secondary flow in cross section 2 (as indicated in the inset) at consecutive instants. Frames are separated by a dimensionless time interval $\Delta t U/d$ of ~0.46

III—Chaotic Flow

For the present curvature $\delta = 0.5$, the flow became chaotic when Re_τ exceeded ~32. As the periodic flow discussed above, also this chaotic flow maintained, in the time averages, the symmetries SU and SM of the base steady flow. Results are shown here for the highest Re_τ investigated (50, yielding Re $\approx$ 324). Figure 8.25 shows contours of the axial velocity (normalized by U) and vector plots of the secondary flow in cross section 2; frames are separated by a time interval of ~0.46d/U. A very high intensity of the velocity fluctuations can be observed, despite the relatively low bulk Reynolds number.

For the same case, Fig. 8.26 reports power spectra of vertical velocity (a) and temperature (b) at the monitoring location "P". Frequency is normalized by U/d. Despite the low bulk Reynolds number (~324), the $-5/3$ power behaviour typical of the inertial subrange of developed turbulence is approximately followed in a relatively broad range of frequencies, and more by temperature than by velocity.

Summarizing, the results for $\delta = 0.5$ indicate the following sequence of flow regimes and bifurcations (by "symmetric" we mean here top-down symmetry about the midplane): steady symmetric base flow $\rightarrow$ (Hopf bifurcation) $\rightarrow$ periodic oscillatory flow, instantaneously anti-symmetric (SU') but with symmetric time mean $\rightarrow$ chaotic flow with symmetric time mean.

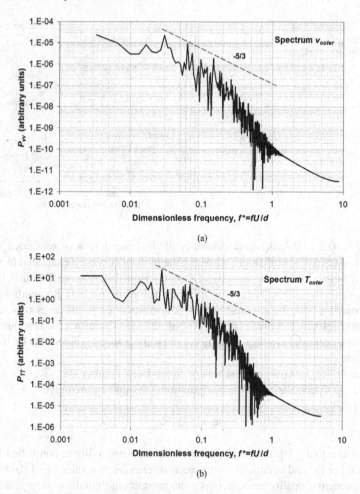

Fig. 8.26 Case $\delta = 0.5$, $Re_\tau = 50$ (chaotic flow with symmetric mean at Re ≈ 324): power spectra at a point located on the equatorial plane in the outer region of section 2. **a** vertical velocity; **b** temperature. Frequency is normalized by U/d

Flow Regime Map

As discussed above, the two curvatures illustrated in detail ($\delta = 0.2$ and $\delta = 0.5$) exhibited a completely different sequence of bifurcations and flow regimes. The other two curvatures examined ($\delta = 0.3$ and $\delta = 0.4$) exhibited yet different sequences (Ciofalo and Di Liberto 2017). In particular, the case $\delta = 0.3$ went through a globally asymmetric periodic flow and the case $\delta = 0.4$ crossed both a steady asymmetric regime, as $\delta = 0.2$, and a periodic asymmetric regime, as $\delta = 0.3$.

Taking account of all the results, a tentative flow regime map can be drawn as in Fig. 8.27. It is represented here in the (Re, δ) plane, but other choices would be possible.

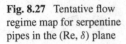

Fig. 8.27 Tentative flow regime map for serpentine pipes in the (Re, δ) plane

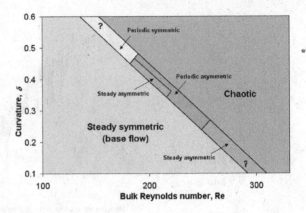

For $\delta = 0.2$ to 0.4, the first instability of the base flow was associated with a sub-critical ("hard") bifurcation, while for the highest curvature examined ($\delta = 0.5$) it was associated with a super-critical ("soft") Hopf bifurcation. The first loss of stability of the base flow occured at different values of the bulk Reynolds number Re, decreasing with the curvature δ, but it corresponded in all cases to values of the friction velocity Reynolds number Re_τ of ~28–30. Therefore, the first instability seems to be controlled by Re_τ (i.e., by the pressure gradient) rather than by Re (i.e., by the flow rate).

However, Re_τ is strongly correlated with the Dean number $De = Re\delta^{1/2}$. A least-squares best-fit correlation of the computational results reported gives

$$De = 1.04\,Re_\tau^{1.41} \tag{8.11}$$

so that it is equally legitimate to state that the first instability is controlled by the Dean number De and occurs when this parameter exceeds a value of ~110–130.

An interesting duality exists between the present configuration (serpentine pipe) and a constant-curvature pipe (e.g., a torus or a helical coil) with the same curvature $\delta = R/c$. An alternating curvature tends to destabilize the base flow, whereas a constant curvature has a stabilizing effect and delays the transition to turbulence, as expressed by Eq. (8.1) reported above in this chapter.

If an alternating curvature pipe (i.e., a serpentine pipe) is conventionally viewed as a pipe with negative curvature, then the results synthesized by Eq. (8.1) and the present computational findings can be reported in a single plot as in Fig. 8.28. Here, symbols represent the present computational results for serpentine pipes ($\delta < 0$) and the consensus value $Re_{cr} = 2100$ for a straight pipe ($\delta = 0$); the continuous line represents Eq. (8.1) for $\delta > 0$, while it is purely qualitative for $\delta < 0$.

Fig. 8.28 Unified stability diagram for pipes with constant and alternating curvature. Symbols represent the results for serpentine pipes ($\delta < 0$); the continuous line represents Eq. (8.1) for $\delta > 0$, while it is purely qualitative for $\delta < 0$

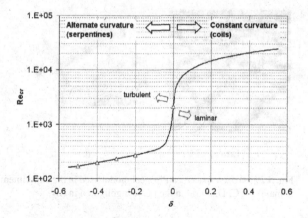

8.4 Transition in Spacer-Filled Plane Channels

Membrane processes, notably Membrane Distillation (MD), often involve flow in flat (plane or spirally wound) channels delimited by selectively permeable membranes. Concentration or temperature polarization reduces the available driving force and thus the efficiency of the process. *Spacers* are widely used to promote mixing and reduce polarization, and serve also the purpose of maintaining a fixed distance between the opposite channel walls (Amokrane et al. 2015). Unfortunately, they also promote pressure drop (Schock and Miquel 1987), so that much research effort, both experimental (Li et al. 2002; Tamburini et al. 2015) and computational (Koutsou et al. 2007; Tamburini et al. 2016) has been devoted to the search for a compromise between high heat/mass transfer and low hydraulic losses.

The author has conducted experiments (based on liquid crystal thermography and digital image processing) and parallel CFD simulations aimed at characterizing spacer-filled channels for membrane distillation. Both overlapped and woven configurations and different values of pitch to height ratio, Reynolds number and flow attack angle were considered (Tamburini et al. 2013, 2016; Ponzio et al. 2017).

The spacers considered here (Fig. 8.29) consist of mutually orthogonal cylindrical filaments of diameter d, arranged in a woven pattern of pitch P.

Figure 8.29a shows the physical spacer model used in the experimental tests, characterized by $d = 5 \times 10^{-3}$ m and $P = 2 \times 10^{-2}$ m. Figure 8.29b shows the unit cell chosen as the computational domain in the numerical simulations, and Fig. 8.29c shows a particular of the computational grid, made up by both hexahedral and tetrahedral volumes. Neglecting the deformation of the filaments at contact points, the filament diameter d coincides with the channel half-height δ so that the channel pitch to height ratio is $P/H = 2$.

Only selected computational results will be discussed here. Dimensionless quantities will be defined with reference to the void (spacerless) configuration, i.e. to a plane channel of height H and thus of hydraulic diameter $D_{eq} = 2H$.

Consistently, the bulk Reynolds number is defined as

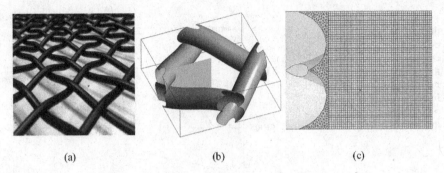

(a)   (b)   (c)

Fig. 8.29 Woven spacers. **a** Physical model used in the experiments; **b** unitary cell (computational domain) for CFD; **c** grid in the plane highlighted in (**b**)

$$\mathrm{Re} = U \times 2H/\nu \tag{8.12}$$

in which $U = Q/(WH)$ is the mean "void channel" streamwise velocity, being Q the volume flow rate and W the channel's spanwise extent (in the experiments, $H = 0.01$ m and $W = 0.22$ m). This definition is preferable to that based on the actual hydraulic diameter and mean velocity of each specific case, and makes comparisons between different configurations more meaningful.

The friction velocity Reynolds number Re_τ is defined as

$$\mathrm{Re}_\tau = u_\tau \delta/\nu \tag{8.13}$$

in which $u_\tau = \tau_w/\rho)^{1/2}$ is the friction velocity. The quantity τ_w is the time- and wall-averaged wall shear stress which, in a void channel under steady-state conditions, would balance the large-scale pressure gradient along the main flow direction s, defined as $f_V = |\partial p/\partial s|$; i.e., one has $\tau_w = \delta f_V$.

The Darcy friction coefficient (four times the Fanning friction factor) is defined as

$$f_D = \frac{4H}{\rho U^2} \left| \frac{\partial p}{\partial s} \right| \tag{8.14}$$

so that, on the basis of the definitions given above for Re and Re_τ, for the plane channel one has $f_D = 128(\mathrm{Re}_\tau/\mathrm{Re})^2$.

The local Nusselt number is defined as

$$Nu = h \frac{2H}{\lambda} \tag{8.15}$$

in which $h = q_w/(T_b - T_w)$ is the local heat transfer coefficient, T_b the fluid bulk temperature and T_w the local wall temperature. The same considerations on averaging the wall heat transfer coefficient developed in the previous Section on serpentine pipes, and leading to the alternative definitions (1) in Eq. (8.5) and (2) in Eq. (8.6) can be applied also to the present spacer-filled channel configuration.

The computational domain was the unit, periodic cell shown in Fig. 8.29b. Mathematically, the problem was described by the continuity, Navier–Stokes and energy equations for a constant-property fluid. A "unit cell" treatment, described e.g. by Tamburini et al. (2016), and similar to that discussed in the previous Sections for the tproidal and serpentine pipes, allowed periodicity conditions to be adopted for all variables at the opposite faces of the computational domain. Thus, the friction Reynolds number Re_τ (i.e., the pressure drop per unit length) was imposed, while the bulk Reynolds number (i.e., the flow rate) was computed as part of the solution.

All simulations were conducted by the finite volume ANSYS-CFX14® code (Ansys Inc 2012). As shown in Fig. 8.29c, it was necessary to adopt a hybrid mesh, with tetrahedral volumes in the regions surrounding the filaments (~30% of the overall volume of the computational domain) and hexahedral volumes in the remaining regions.

No slip conditions ($u_i = 0$) were imposed at the walls and on the filaments' surface. As regards the thermal boundary conditions, the bottom wall and the filaments were assumed to be adiabatic ($q_w = 0$), while at the top (thermally active) wall a Robin (3rd type, or mixed) condition was imposed:

$$T_w - T_c = -r_{ext}q_w \qquad (8.16)$$

mimicking the experimental boundary condition. Values representative of those holding in the experiments were chosen for the outer cold temperature T_c and the thermal resistance r_{ext} (19 °C and 6.24×10^{-3} m²K/W, respectively). Also the spacer sizes were the same as the experimental ones and the fluid was assumed to be water at 39 °C with a Prandtl number of 4.63. Finally, as in the experiments, two orientations of the main flow with respect to the spacer filaments of the upper layer were considered (0° and 45°). In the present study, all simulations were run in time-dependent mode and attained either a steady-state or a time-dependent (time periodic or irregular, i.e. chaotic) regime.

Results for $\vartheta = 0$–$90°$

For this orientation (main flow parallel and orthogonal to the two filament layers), steady-state flow was predicted for Re_τ up to ~ 105 (Re up to ~340), periodic flow in a narrow interval about $Re_\tau \approx 10$ (Re ≈ 350) and chaotic flow for $Re_\tau \approx 120$ or larger (Re $\geq$ ~390).

For the sake of brevity, steady-state results are not reported. As an example of a complex but still time-periodic solution, Fig. 8.30a reports a time series of the streamwise velocity u_s at the centre of the unit cell for $Re_\tau = 110$, or Re ≈ 352 (time is normalized by δ/U, velocity by the mean streamwise velocity U). The time interval shown includes three periods. The duration t_{per}[1] of each period is 101.15 δ/U, corresponding to ~42 s at the scale of the experiments.

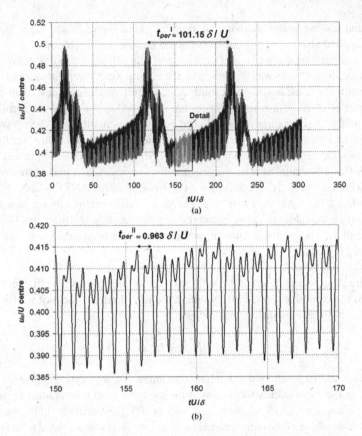

Fig. 8.30 Time series of the streamwise velocity u_s at the centre of the unit cell for $\vartheta = 0$–$90°$ and $Re_\tau = 110$ (Re $= 352$, periodic flow). Velocity is normalized by the mean streamwise velocity U, time by δ/U. **a** Large scale behaviour over a time interval including three periods $t_{per}{}^I$. **b** Detail of the behaviour of u_s in the interval highlighted in (**a**), showing a secondary "quasi-period" of $t_{per}{}^I/105$

Graph (b) reports an enhanced detail of the behaviour of u_s in the shorter interval highlighted in (a), showing a secondary "quasi-period" $t_{per}^I/105$ which corresponds to $0.963 \times \delta/U$, or ~0.4 s at the scale of the experiments.

The corresponding power spectrum is reported in Fig. 8.31; here, frequency is normalized as $F = f/(U/\delta)$, i.e. as a Strouhal number based on mean velocity and filament diameter, while P_{uu} is in arbitrary units. Discrete peaks, corresponding to a base dimensionless frequency F^I of ~0.01 and its harmonics, are visible; the very high order harmonic at $105\,F^I$ corresponds to the rapid oscillations in Fig. 8.30(b).

An example of chaotic behaviour is illustrated in Fig. 8.32, which reports computational results for $\vartheta = 0$–$90°$ and $Re_\tau = 140$ (Re $= 478$). Graph (a) shows a time series of the streamwise velocity u_s at the centre of the unit cell, graph (b) the corresponding time power spectrum. Quantities are normalized as in Figs. 8.30 and 8.31. Velocity

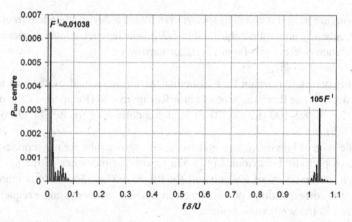

Fig. 8.31 Power spectrum of the streamwise velocity u_s at the centre of the unit cell for $\vartheta = 0$–$90°$ and $Re_\tau = 110$ (Re = 352), reported as a function of time in Fig. 8.30 (frequency is normalized by U/δ, i.e. as a Strouhal number, while P_{uu} is in arbitrary units)

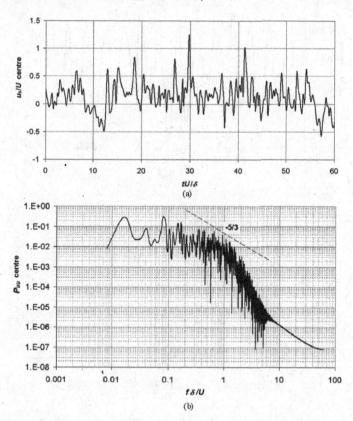

Fig. 8.32 Computational results for $\vartheta = 0$–$90°$ and $Re_\tau = 140$ (Re = 478, chaotic flow). **a** Streamwise velocity u_s at the centre of the unit cell. **b** Corresponding power spectrum. Quantities are normalized as in Figs. 8.30 and 8.31

fluctuations are irregular and their spectrum includes many independent frequencies, but the frequency interval exhibiting the -5/3 power law behaviour characteristic of full turbulence (inertial sub-range) is rather narrow.

Results for $\vartheta = 45°$

For this orientation (main flow bisecting the angle formed by the two filament layers), steady-state flow was predicted for Re_τ up to ~70 (Re up to ~ 225), periodic flow for $Re_\tau \approx 80–100$ (Re $\approx$ 270–365) and chaotic flow for $Re_\tau \approx 120$ or larger (Re $\geq$ ~465).

For the sake of brevity, representative steady-state results are not reported. As an example of periodic behaviour, Fig. 8.33a reports a time series of the streamwise velocity u_s at the centre of the unit cell for $Re_\tau = 100$, or Re $\approx$ 365. Figure 8.33b reports the corresponding power spectrum. Time, velocity and frequency are normalized as in the previous figures.

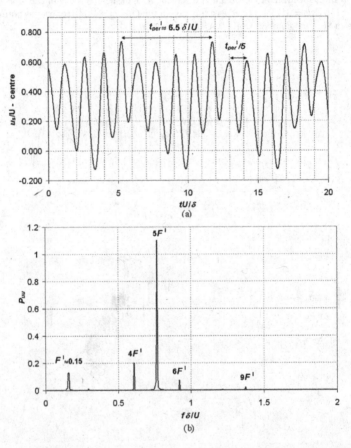

Fig. 8.33 Computational results for $\vartheta = 45°$ and $Re_\tau = 100$ (Re = 365, periodic flow). **a** Behaviour of the streamwise velocity u_s at the centre of the unit cell. **b** Corresponding power spectrum. Quantities are normalized as in Figs. 8.30 and 8.31

The time interval considered in Fig. 8.33 includes three periods, each of duration $t_{per}{}^{\mathrm{I}} \approx 6.5\ \delta/U$ (corresponding to ~2.65 s at the scale of the experiments). In the spectral density diagram, this periodicity corresponds to the base frequency $F^{\mathrm{I}} \approx 0.15$. However, most of the fluctuation intensity is associated with a secondary period $t_{per}{}^{\mathrm{II}} = t_{per}{}^{\mathrm{I}}/5$ (i.e., ~1.3δ/U, or ~0.53 s at the scale of the experiments), corresponding to the spectral peak at $F^{\mathrm{II}} = 5F^{\mathrm{I}}$. Harmonics at 4, 6 and $9 \times F^{\mathrm{I}}$ are also appreciable.

Finally, an example of fully chaotic behaviour is reported also for the 45° orientation in Fig. 8.34, which is for $\mathrm{Re}_\tau = 120$ ($\mathrm{Re} = 465$). The quantities reported are the same as in Fig. 8.32, and time series and spectra are similar. In this case, the power spectrum P_{uu} in Fig. 8.34b exhibits a marked peak at $f\,\delta/U \approx 2.3$ which does not have a clear counterpart in the case $\vartheta = 0\text{–}90°$ of Fig. 8.32b.

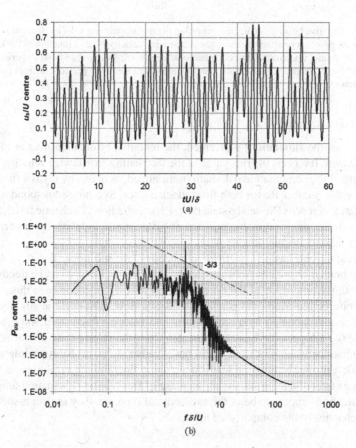

Fig. 8.34 Computational results for $\vartheta = 45°$ and $\mathrm{Re}_\tau = 120$ ($\mathrm{Re} = 465$, chaotic flow). **a** Streamwise velocity u_s at the centre of the unit cell. **b** Corresponding power spectrum. Quantities are normalized as in Figs. 8.30 and 8.31

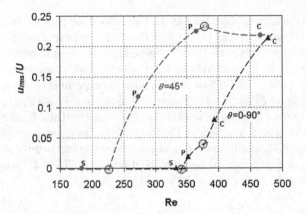

Fig. 8.35 Tentative bifurcation diagram reporting the rms streamwise velocity fluctuation $u_{s,rms}$ at a monitoring point (normalized by the mean streamwise velocity U) as a function of the Reynolds number Re. Symbols correspond to computational results (S: steady-state flow; P: periodic flow; C: chaotic flow). Shaded circles indicate bifurcations from steady to periodic or from periodic to chaotic flow

Bifurcation Diagram

Based on all the simulations performed, the tentative bifurcation map in Fig. 8.35 can be drawn. It reports the rms value of the fluctuating streamwise velocity, u_{rms}, at the centre of the computational domain (normalized, as usual, by U) as a function of the Reynolds number Re for both flow attack angles. Symbols correspond to actual computational results (S: steady-state flow; P: periodic flow; C: chaotic flow). Shaded circles indicate the presumable position of bifurcations from steady to periodic or from periodic to chaotic flow.

Of course, one has $u_{rms} = 0$ for all steady-state solutions. For $\vartheta = 45°$, the periodic branch bifurcates from the steady-state one at Re≈225 and extends above Re $= 350$; the amplitude of the fluctuations increases considerably in this range. A second bifurcation from time-periodic to chaotic flow must occur at slightly higher Re, and is accompanied by a reduction of the velocity fluctuation amplitude. For $\vartheta = 0$–90°, the periodic branch bifurcates from the steady-state one at some Re intermediate between ~330 and ~ 350 but is rather short, probably ending at Re $=$ 370–380; a further bifurcation leads to a chaotic flow branch exhibiting, at least in the range studied here, a monotonic increase of the velocity fluctuation amplitude. All bifurcations appear to be of the supercritical type, i.e. they are not associated to abrupt changes in the computed quantities.

References

Amokrane M, Sadaoui D, Koutsou CP, Karabelas AJ, Dudeck M (2015) A study of flow field and concentration polarization evolution in membrane channels with two-dimensional spacers during water desalination. J Membr Sci 477:139–150

Ansys Inc (2012) Ansys-CFX reference guide release 14 5

Berger SA, Talbot L, Yao LS (1983) Flow in curved pipes. Annu Rev Fluid Mech 15:461–512

Boussinesq MJ (1868) Mémoire sur l'influence des frottements dans les mouvements régulier des fluids. Journal de Mathématiques Pures et Appliquées 2me Série 13:377–424

Ciofalo M, Di Liberto M (2015) Fully developed laminar flow and heat transfer in serpentine pipes. Int J Thermal Sci 96:248–266

Ciofalo M, Di Liberto M (2017) Transition to turbulence in serpentine pipes. Int J Thermal Sci 116:129–149

Ciofalo M, Di Liberto M, Di Piazza I (2012) Steady, periodic, quasi-periodic and chaotic flow regimes in toroidal pipes. In: Lazzari S, Rossi di Schio E (eds) Proceedings of the 30th UIT heat transfer conference, Bologna, Italy, 25–27 June 2012. Esculapio, Bologna, pp 5–16

Cioncolini A, Santini L (2006) An experimental investigation regarding the laminar to turbulent flow transition in helically coiled pipes. Exp Thermal Fluid Sci 30:367–380

Dean WR (1927) Note on the motion of the fluid in a curved pipe. Phil Mag 4:208–223

Di Liberto M, Ciofalo M (2013) Turbulence structure and budgets in curved pipes. Comput Fluids 88:452–472

Di Piazza I, Ciofalo M (2010) Numerical prediction of turbulent flow and heat transfer in helically coiled pipes. Int J Thermal Sci 49:653–663

Di Piazza I, Ciofalo M (2011) Transition to turbulence in toroidal pipes. J Fluid Mech 687:72–117

Fenstermacher PR, Swinney HL, Gollub JP (1979) Dynamic instability and the transition to chaotic Taylor vortex flow. J Fluid Mech 94:103–128

Isachenko V, Osipova V, Sukomel A (1974) Heat transfer, 2nd edn. Mir Publishers, Moscow

Ito H (1959) Friction factors for turbulent flow in curved pipes. ASME J Basic Eng 81:123–134

Janssen LAM, Hoogendoorn CJ (1978) Laminar convective heat transfer in helical coiled tubes. Int J Heat Mass Transf 21:1197–1206

Koutsou CP, Yiantsios SG, Karabelas AJ (2007) Direct numerical simulation of flow in spacer-filled channels: effect of spacer geometrical characteristics. J Membr Sci 291:53–69

Kuznetsov Y (2004) Elements of applied bifurcation theory. Springer, New York

Li F, Meindersma W, de Haan AB, Reith T (2002) Optimization of commercial net spacers in spiral wound membrane modules. J Membr Sci 208:289–302

Ponzio F, Tamburini A, Cipollina A, Micale G, Ciofalo M (2017) Experimental and computational investigation of heat transfer in channels filled by woven spacers. Int J Heat Mass Transf 104:163–177

Ruelle D, Takens F (1971) On the nature of turbulence. Commun Math Phys 20:167–192

Schock G, Miquel A (1987) Mass transfer and pressure loss in spiral wound modules. Desalination 64:339–352

Sobey IJ, Drazin PG (1986) Bifurcations of two-dimensional channel flows. J Fluid Mech 171:263–287

Sreenivasan KR, Strykowski PJ (1983) Stabilization effects in flow through helically coiled pipes. Exp Fluids 1:31–36

Tamburini A, Pitò P, Cipollina A, Micale G, Ciofalo M (2013) A Thermochromic liquid crystals image analysis technique to investigate temperature polarization in spacer-filled channels for membrane distillation. J Membr Sci 447:260–273

Tamburini A, Cipollina A, Al-Sharif S, Albeyrutty M, Gurreri L, Micale G, Ciofalo M (2015) Assessment of temperature polarization in membrane distillation channels by liquid crystal thermography. Desal Water Treat 55:2747–2765

Tamburini A, Renda M, Cipollina A, Micale G, Ciofalo M (2016) Investigation of heat transfer in spacer-filled channels by experiments and direct numerical simulations. Int J Heat Mass Transf 93:1190–1205

Vashisth S, Kumar V, Nigam KDP (2008) A review on the potential applications of curved geometries in process industry. Ind Eng Chem Res 47:3291–3337

Webster DR, Humphrey JAC (1993) Experimental observations of flow instability in a helical coil. J Fluids Eng 115:436–443

Webster DR, Humphrey JAC (1997) Travelling wave instability in helical coil flow. Phys Fluids 9:407–418

Woitkowiak J, Popiel CO (2000) Effect of cooling on pressure losses in U-type wavy pipe flow. Int Commun Heat Mass Transf 27:169–177

Xin RC, Ebadian MA (1997) The effects of Prandtl numbers on local and average convective heat transfer characteristics on helical pipes. Int J Heat Mass Transf 119:467–473

Conclusions

God keep me from ever completing anything. This whole book is but a draught – nay, but the draught of a draught. Oh, Time, Strength, Cash, and Patience!

Hermann Melville, Moby Dick, Chapter 32

The overview of models and applications presented in this work can barely provide an idea of the immense theoretical, computational and experimental efforts dedicated to the study of turbulence so far.

Regretfully, whole important subjects have remained out of the present work, e.g. turbulence in compressible and multi-phase fluids, turbulent reactive flows (including combustion) and experimental techniques in turbulence studies. Other subjects, like scalar transport and mixing, have just been touched in a concise way.

On the other hand, some subjects that are rarely given space in similar treatises, including buoyant, transient and transitional turbulence, have received some attention here.

In extreme synthesis, methods for the numerical simulation of turbulence can be classified into the three great families of Direct Numerical Simulation, or DNS; Reynolds-Averaged Navier–Stokes simulation, or RANS, based on time-, phase- or ensemble-averaging; and Large Eddy Simulation, or LES, based on spatial filtering.

Direct Numerical Simulation is invaluable as a tool to elucidate the fundamental mechanisms of turbulence and is increasingly replacing experimental investigations, on which it offers the great advantage of a complete and ordered access to the whole flow field. However, despite the impressive increase in computing power, made possible mainly by large-scale parallelism (see Sect. 4.3), DNS is hard to perform in high Reynolds number flows and complex geometries, requires a heavy and sophisticated post-processing of the results, and thus is mainly confined, at least for the time being, to research applications.

RANS models, the first to be developed, remain the most widespread simulation tool in all industrial applications, in which the quantities of interest are mean values, extended parametric studies are often to be conducted, and geometrical complexity,

M. Ciofalo, *Thermofluid Dynamics of Turbulent Flows*, UNIPA Springer Series, https://doi.org/10.1007/978-3-030-81078-8

or the simultaneous presence of different physical phenomena (combustion, radiative heat transfer, chemical reactions, multiphase flow) make other approaches excessively cumbersome and costly.

In the context of RANS closure models, it can probably be stated that the great expectations raised in the years 1980–1990 by differential stress models (DSM), based on separate transport equations for the Reynolds stresses/fluxes, have largely been disappointed. These models remain computationally expensive and unstable, require a large number of calibration "constants", and often miss their declared goal, i.e. the accurate prediction of flow and heat transfer quantities in problems characterized by a strong anisotropy of the normal turbulent stresses. At the same time, many of the reservations initially expressed by the turbulence community on eddy viscosity models have largely been superseded. The main reasons are:

- several shortcomings of predictions based on eddy viscosity models, evidenced in the early years of their development, have turned out to be caused more by an insufficient spatial resolution (then almost inevitable) than by intrinsic limits of the models, and have significantly diminished as fine computational grids, accurate higher order discretization methods, and an adequate number of iterations have become routinary;
- new eddy viscosity models have been introduced, such as RNG $k - \varepsilon$, $k - \omega$ and SST, that overcome many of the well known shortcomings of early models (e.g. $k - \varepsilon$) and provide accurate results even in situations initially considered untractable, such as recirculating flows with separation and reattachment, natural or mixed turbulent convection, intermediate and transitional regimes or transient turbulence.

Therefore, it can be expected that eddy viscosity models still have a bright future, in one or another of their many variants and reincarnations.

Finally, Large Eddy Simulation (LES) has come out of its pioneering stage since the years 1980 and has amply proved to possess the ability to become a valid predictive tool for complex thermofluid dynamics problems. Advanced CFD codes such as Fluent, CFX and STAR-CD currently include LES as a standard turbulence modelling option.

Among subgrid models for LES, probably the best results have been provided by the "dynamic" one, combining the simplicity and stability of eddy viscosity models with a great generality and the almost complete freedom from empirical parameters. The hybrid DES (Detached Eddy Simulation) approach, which combines LES in the free stream with RANS in the near-wall region, is also increasingly adopted.

As a conclusive remark, it should be stressed that any turbulence model has the function of predicting the *effects* that turbulence exerts on the mean flow, rather than predicting the features of turbulence itself, and should be judged for its ability to fulfill this, and only this, purpose. The only rigorous approach to the prediction of turbulent fluctuations, their moments and their spatial and temporal structure, is the direct numerical simulation of turbulence.

Appendix
Elements of Tensor Algebra

Note: In the following, vectors and tensors will be expressed in a *Cartesian orthogonal* reference frame (O, x_1, x_2, x_3), or (O, x, y, z). In such a frame, the distinction between covariant and contravariant components vanishes, with great benefits for simplicity of notation.

A.1 Vector and Tensor Notation

Scalar φ (e.g. temperature T, concentration C, turbulent kinetic energy k ...):

$$\text{In } both \text{ notations, it is denoted by a simple symbol with no subscripts:} \varphi \qquad (A.1)$$

- *Vector* **a** (e.g. velocity **u**, acceleration **g** ...):

$$\text{Vector notation (using bold characters)} : \quad \mathbf{a} \qquad (A.2)$$

$$\text{Tensor notation (using subscripts)} : \quad a_i, i = 1, 2, 3 \qquad (A.3)$$

- *Gradient of a scalar* φ (=vector):

$$\text{Vector notation} : \quad \nabla\varphi = \frac{\partial\varphi}{\partial x}\mathbf{i} + \frac{\partial\varphi}{\partial y}\mathbf{j} + \frac{\partial\varphi}{\partial z}\mathbf{k} \qquad (A.4)$$

$$\text{Tensor notation} : \quad \frac{\partial\varphi}{\partial x_i} \qquad (A.5)$$

in which ∇ (Nabla) is the symbolic vector of components $(\partial/\partial x, \partial/\partial y, \partial/\partial z)$.

© The Editor(s) (if applicable) and The Author(s), under exclusive license to Springer Nature Switzerland AG 2022
M. Ciofalo, *Thermofluid Dynamics of Turbulent Flows*, UNIPA Springer Series, https://doi.org/10.1007/978-3-030-81078-8

- *Scalar product of two vectors* **a**, **b** (=scalar):

$$\text{Vector notation :} \quad \mathbf{a} \cdot \mathbf{b} \tag{A.6}$$

$$\text{Tensor notation :} \quad a_j b_j, \text{ shorthand for } \sum_{j=1}^{3} a_j b_j \tag{A.7}$$

(Note the use of *Einstein's convention* of implicit summation over repeated indices. The "*j*" index is *dummy*, i.e., it could be replaced by any other symbol).

- *Divergence of a vector* **a** (=scalar), which can be expressed as the scalar product of the symbolic vector $\nabla = (\partial/\partial x, \partial/\partial y, \partial/\partial z)$ (Nabla) by **a**:

$$\text{Vector notation :} \quad \nabla \cdot \mathbf{a} = \frac{\partial a_x}{\partial x} + \frac{\partial a_y}{\partial y} + \frac{\partial a_z}{\partial z} \tag{A.8}$$

$$\text{Tensor notation :} \quad \frac{\partial a_j}{\partial x_j}, \text{ shorthand for } \sum_{j=1}^{3} \frac{\partial a_j}{\partial x_j} \tag{A.9}$$

- *Vector product* (or *external product*) *of two vectors* **a**, **b** (=vector):

Vector notation :

$$\mathbf{a} \times \mathbf{b} = \begin{vmatrix} \mathbf{i} & \mathbf{j} & \mathbf{k} \\ a_x & a_y & a_z \\ b_x & b_y & b_z \end{vmatrix} = (a_y b_z - a_z b_y)\mathbf{i} + (a_z b_x - a_x b_z)\mathbf{j} + (a_x b_y - a_y b_x)\mathbf{k} \tag{A.10}$$

$$\text{Tensor notation :} \quad \varepsilon_{ijk} a_j b_k, \text{ shorthand for } \sum_{j=1}^{3} \sum_{k=1}^{3} \varepsilon_{ijk} a_j b_k \tag{A.11}$$

in which ε_{ijk} is the third order *alternating tensor*, defined by $\varepsilon_{ijk} = 1$ if the subscripts i, j, k form an even permutation, $\varepsilon_{ijk} = -1$ if i, j, k form an odd permutation, $\varepsilon_{ijk} = 0$ if two of the subscripts are equal.

- *Curl of a vector* **a** (vector), which can be expressed as the external product of the symbolic vector $\nabla = (\partial/\partial x, \partial/\partial y, \partial/\partial z)$ (Nabla) by **a**:

$$\text{Vector notation :} \quad \nabla \times a = \begin{vmatrix} \mathbf{i} & \mathbf{j} & \mathbf{k} \\ \frac{\partial}{\partial x} & \frac{\partial}{\partial y} & \frac{\partial}{\partial z} \\ a_x & a_y & a_z \end{vmatrix}$$

$$= \left(\frac{\partial a_z}{\partial y} - \frac{\partial a_y}{\partial z} \right)\mathbf{i} + \left(\frac{\partial a_x}{\partial z} - \frac{\partial a_z}{\partial x} \right)\mathbf{j} + \left(\frac{\partial a_y}{\partial x} - \frac{\partial a_x}{\partial y} \right)\mathbf{k} \tag{A.12}$$

$$\text{Tensor notation}: \varepsilon_{ijk}\frac{\partial a_j}{\partial x_k}, \text{ shorthand for } \sum_{j=1}^{3}\sum_{k=1}^{3}\varepsilon_{ijk}\frac{\partial a_j}{\partial x_k} \tag{A.13}$$

- *Second-order tensor* (e.g. diffusivity tensor Γ, stress tensor $\mathbb{P}$)

$$\text{Vector notation: } \Gamma \tag{A.14}$$

$$\text{Tensor notation: } \Gamma_{ij} \tag{A.15}$$

- *Product of a second-order tensor Γ by a vector $\mathbf{a}$ (=vector):*

$$\text{Vector notation}: \quad \Gamma \cdot \mathbf{a} \tag{A.16}$$

$$\text{Tensor notation}: \quad \Gamma_{ij}a_j, \text{ shorthand for } \sum_{j=1}^{3}\Gamma_{ij}a_j \tag{A.17}$$

Note that this product follows the rule of matrix-by-vector multiplication.

A.2   Invariants of a Second-Order Tensor

These are scalar quantities that can be built from the components of a tensor and do not change under coordinate transformations. Three such invariants can be defined:

Linear invariant, or *trace* : $I_1 = \text{tr}(\Gamma) = \Gamma_{11} + \Gamma_{22} + \Gamma_{33} = \Gamma_{kk}$ (implicit sum)
$$\tag{A.18}$$

In a second-order tensor, I_1 is the sum of the three diagonal components. If $1/3$ of the trace is subtracted from each diagonal component, a tensor Γ' whose trace is zero (a *traceless* tensor) is obtained:

$$\Gamma'_{ij} = \Gamma_{ij} - \frac{1}{3}\delta_{ij}\Gamma_{kk} \text{ with } \text{tr}(\Gamma') = 0 \tag{A.19}$$

Here, δ_{ij} is the *Kronecker delta* (itself a second-order tensor), defined by

$$\begin{aligned}\delta_{ij} &= 1 \quad \text{if } i = j \\ \delta_{ij} &= 0 \quad \text{if } i \neq j\end{aligned} \tag{A.20}$$

$$\textit{Quadratic invariant}: \quad I_2 = \Gamma_{kk}\Gamma_{kk} \text{(implicit sum)} \tag{A.21}$$

In a second-order tensor, I_2 is the sum of the squares of all 9 components.

$$Cubic\,invariant: \quad I_3 = \det(\Gamma_{ij}) \tag{A.22}$$

In a second-order tensor, I_3 is the determinant of the 3×3 matrix formed by the tensor's components (note that such matrix is *not* the tensor, but one of its possible representations).

A.3   Kinematic Tensors

$$Strain-rate \text{ tensor } \mathbb{S}: \quad S_{ij} = \frac{1}{2}\left(\frac{\partial u_i}{\partial x_j} + \frac{\partial u_j}{\partial x_i}\right) \tag{A.23}$$

Note that S_{ij} is *symmetric* ($S_{ij} = S_{ji}$) and that $S_{kk} = \frac{\partial u_j}{\partial x_j}$ (implicit sum), i.e. $\text{tr}(\mathbb{S}) = \nabla \cdot \mathbf{u}$.

$$Vorticity \text{ tensor, or } rotation \text{ tensor } \Omega \quad \Omega_{ij} = \frac{1}{2}\left(\frac{\partial u_i}{\partial x_j} - \frac{\partial u_j}{\partial x_i}\right) \tag{A.24}$$

Note that Ω_{ij} is *anti-symmetric* ($\Omega_{ij} = -\Omega_{ij}$). Note also the possible decomposition of the velocity derivatives as the sum of symmetric and anti-symmetric parts:

$$\frac{\partial u_i}{\partial x_j} = S_{ij} + \Omega_{ij} \tag{A.25}$$

Vorticity can also be defined as a *vector*, equal to ½ the curl of velocity:

$$\omega = \frac{1}{2}\nabla \times \mathbf{u} \quad \text{(in vector notation)} \tag{A.26}$$

$$\omega_i = \frac{1}{2}\varepsilon_{ijk}\frac{\partial u_j}{\partial x_k} \quad \text{(in tensor notation)} \tag{A.27}$$

It is easily proved that

$$\omega_1 = \Omega_{32} = -\Omega_{23} \quad \omega_2 = \Omega_{13} = -\Omega_{31} \quad \omega_3 = \Omega_{21} = -\Omega_{12} \tag{A.28}$$

In plane flows ($\partial/\partial z = 0$, $w = 0$) the vector ω reduces to $(0, 0, \omega_z)$ and the only nonzero component ω_z can be treated as a simple scalar ω; this is commonly done in the popular stream function-vorticity treatment of planar incompressible flows.

Mainly for flow visualization purposes, a scalar called *Okubo-Weiss parameter*, or *Q-parameter*, is often defined:

$$Q = \omega_{ij}\omega_{ij} - S_{ij}S_{ij} \tag{A.29}$$

Q is positive in regions where vorticity prevails over the strain rate, and negative where the opposite happens (*irrotational strain* regions), while it vanishes in the proximity of solid walls.

A.4 Stresses and Constitutive Equations

A.4.1 Stress Tensor

The stress tensor $\mathbb{P}$, or $\wp_{ij}$, is a second order tensor. With reference to Figure A.1, it can be regarded as an *operator* which, when applied to the versor $\mathbf{n}$ normal to an oriented surface element $\mathbf{dS} = dS\,\mathbf{n}$, yields the stress $\mathbf{f}$ (force per unit surface area) acting on that surface element:

$$\mathbf{f} = \mathbb{P} \cdot \mathbf{n} \quad \text{(in vector notation)} \tag{A.30}$$

$$f_i = \wp_{ij} n_j \quad \text{(in tensor notation)} \tag{A.31}$$

This definition holds for any continuum (thus, not only for fluids but also for solids). The individual components $\wp_{ij}$ of the stress tensor can be given an important physical interpretation. In fact, consider an oriented surface element parallel to one of the Cartesian planes (say, $x_3 x_1 = zx$) and facing the positive direction of the orthogonal ($x_2 = y$) axis. The associated versor $\mathbf{n}^{(2)}$ coincides with the versor $\mathbf{j}$ of the y axis and its components are (0 1, 0). Therefore, the stress acting on the face zx is the vector

$$\mathbf{f}^{(2)} = \begin{pmatrix} \wp_{11} & \wp_{12} & \wp_{13} \\ \wp_{21} & \wp_{22} & \wp_{23} \\ \wp_{31} & \wp_{32} & \wp_{33} \end{pmatrix} \begin{pmatrix} 0 \\ 1 \\ 0 \end{pmatrix} = \begin{pmatrix} \wp_{12} \\ \wp_{22} \\ \wp_{32} \end{pmatrix} \tag{A.32}$$

Fig. A.1 Stress $\mathbf{f}$ acting on a surface element dS of versor $\mathbf{n}$

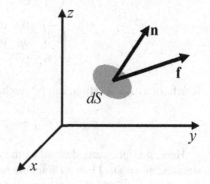

i.e., the second column of the matrix representing the tensor $\mathbb{P}$. More generally:

- the j-th column of the tensor $\mathbb{P}$ represents the stress vector acting on the oriented surface element orthogonal to the direction j;
- the generic element $\wp_{ij}$ of the tensor $\mathbb{P}$ is the i-th component of the stress acting on the oriented surface element orthogonal to the direction j.

Therefore, the stress tensor can be regarded not only as an operator, but also as the *ordered collection of the stress components* acting on oriented surface elements orthogonal to the coordinate axes.

The diagonal elements $\wp_{ii}$ of the stress tensor represent *normal* stresses, while the elements out of the diagonal, $\wp_{ij}$ ($i \neq j$), represent *tangential* stresses.

To preserve equilibrium with respect to rotation, it is necessary that, in the absence of concentrated forces, $\wp_{ij} = \wp_{ji}$, i.e. the stress tensor is *symmetric*.

A.4.2 Constitutive Equations

Whatever the nature of the continuum under consideration, *constitutive equations* establish a link between the stress components $\wp_{ij}$ and some other quantities. In *solids*, these latter quantities usually are the *deformations* (strains) with respect to a reference configuration. In fluids, where strains themselves are devoid of physical significance, the link must rather be sought with the *flow field*, and in particular with the time derivatives of strains (*strain rates*), which coincide with the spatial derivatives of velocities.

In fluids, the normal and tangential stress components have a quite different physical meaning and exhibit a completely different dependence upon the flow field; suffice it to consider that in a still fluid pressure is still present, whereas tangential stresses vanish. Therefore, it is customary to *decompose* the stress tensor as

$$\begin{pmatrix} \wp_{11} & \wp_{12} & \wp_{13} \\ \wp_{21} & \wp_{22} & \wp_{23} \\ \wp_{31} & \wp_{32} & \wp_{33} \end{pmatrix} = \begin{pmatrix} -p & 0 & 0 \\ 0 & -p & 0 \\ 0 & 0 & -p \end{pmatrix} + \begin{pmatrix} \wp_{11}+p & \wp_{12} & \wp_{13} \\ \wp_{21} & \wp_{22}+p & \wp_{23} \\ \wp_{31} & \wp_{32} & \wp_{33}+p \end{pmatrix}$$

$$= \begin{pmatrix} -p & 0 & 0 \\ 0 & -p & 0 \\ 0 & 0 & -p \end{pmatrix} + \begin{pmatrix} \tau_{11} & \tau_{12} & \tau_{13} \\ \tau_{21} & \tau_{22} & \tau_{23} \\ \tau_{31} & \tau_{32} & \tau_{33} \end{pmatrix} \qquad (A.33)$$

which, in tensor notation, can be synthetically written as

$$\wp_{ij} = -p\delta_{ij} + (\wp_{ij} + p\delta_{ij}) = -p\delta_{ij} + \tau_{ij} \qquad (A.34)$$

Here, p is pressure, defined as the opposite of the average of the three normal stresses, $p = \wp_{kk}/3$ (with implicit summation). The tensor $\tau_{ij} = \wp_{ij} + p\delta_{ij} = \wp_{ij} - 1/3\delta_{ij}\wp_{kk}$ is the *deviatoric part* of the stress tensor and can be denoted as *viscous stress tensor*. By construction, it is *traceless* ($\tau_{kk} = 0$). Note that, in general fluids, viscous stresses are *mainly* tangential, but small normal viscous stresses $\tau_{ij} = \wp_{ii} + p$ (no summation) may exist.

In fluids called *Newtonian*, $\wp_{11} = \wp_{22} = \wp_{33} = -p$, so that τ_{ij} has zero diagonal, i.e. it is purely tangential (normal viscous stresses are nil). In addition, in Newtonian fluids the traceless part of the stress tensor can be assumed to be proportional to the corresponding traceless part of the strain rate tensor S_{ij}:

$$(\wp_{ij} - 1/3\delta_{ij}\wp_{kk}) = 2\mu(S_{ij} - 1/3\delta_{ij}S_{kk}) \tag{A.35}$$

in which μ is the fluid's viscosity. This assumption is called "Boussinesq hypothesis for viscous stresses". Since $\wp_{ij} - 1/3\delta_{ij}\wp_{kk} = \tau_{ij}$ and $S_{kk} = \nabla\mathbf{u}$ (velocity divergence), it can be written

$$\tau_{ij} = 2\mu(S_{ij} - 1/3\delta_{ij}\nabla\mathbf{u}) \tag{A.36}$$

and, for constant-density fluids ($\nabla\mathbf{u} = 0$),

$$\tau_{ij} = 2\mu S_{ij} = \mu(\partial u_i/\partial y_j + \partial u_j/\partial y_i) \tag{A.37}$$

This last expression is the three-dimensional generalization of the well known "Newton's law of viscosity" $\tau = \mu\ \partial u/\partial y$, which expresses the tangential viscous stress as proportional to the velocity gradient in the simple case in which velocity has only one nonzero component (say, along x) and this is a function of only one coordinate orthogonal to x (say, y).

Printed in the United States
by Baker & Taylor Publisher Services